The Gluon

Edited by Paul F. Kisak

Contents

Chapter 1

Gluon

Gluons /ˈgluːɒnz/ are elementary particles that act as the exchange particles (or gauge bosons) for the strong force between quarks, analogous to the exchange of photons in the electromagnetic force between two charged particles.[6]

In technical terms, gluons are vector gauge bosons that mediate strong interactions of quarks in quantum chromodynamics (QCD). Gluons themselves carry the color charge of the strong interaction. This is unlike the photon, which mediates the electromagnetic interaction but lacks an electric charge. Gluons therefore participate in the strong interaction in addition to mediating it, making QCD significantly harder to analyze than QED (quantum electrodynamics).

1.1 Properties

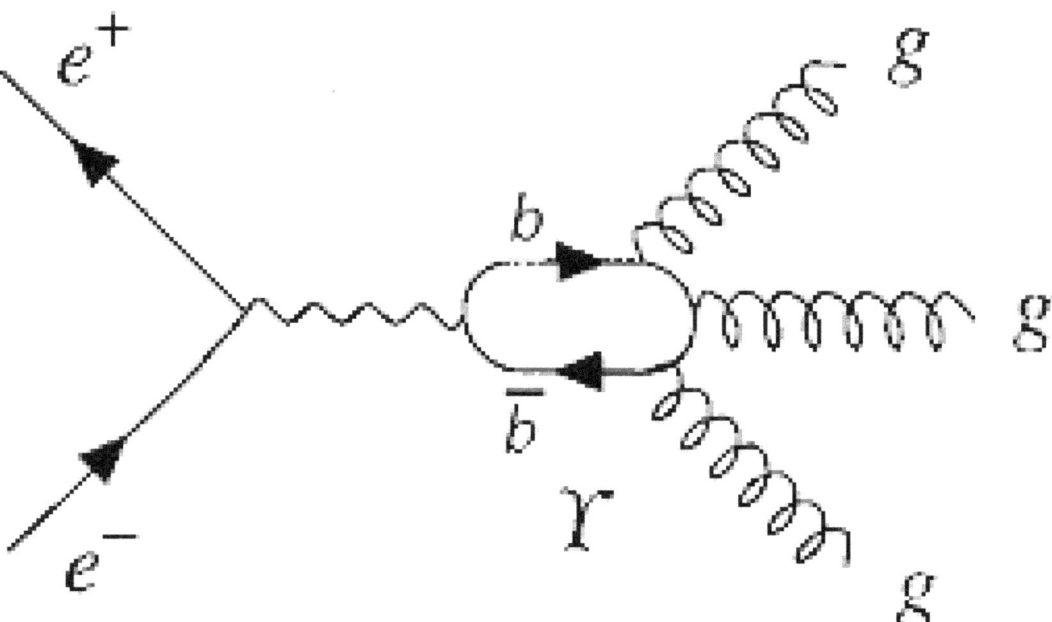

Diagram 2: $e^+e^- \rightarrow \Upsilon(9.46) \rightarrow 3g$

The gluon is a vector boson; like the photon, it has a spin of 1. While massive spin-1 particles have three polarization states, massless gauge bosons like the gluon have only two polarization states because gauge invariance requires the polarization to be transverse. In quantum field theory, unbroken gauge invariance requires that gauge bosons have zero mass (experiment limits the gluon's rest mass to less than a few meV/c^2). The gluon has negative intrinsic parity.

1.2   Numerology of gluons

Unlike the single photon of QED or the three W and Z bosons of the weak interaction, there are eight independent types of gluon in QCD.

This may be difficult to understand intuitively. Quarks carry three types of color charge; antiquarks carry three types of anticolor. Gluons may be thought of as carrying both color and anticolor, but to correctly understand how they are combined, it is necessary to consider the mathematics of color charge in more detail.

1.2.1   Color charge and superposition

In quantum mechanics, the states of particles may be added according to the principle of superposition; that is, they may be in a "combined state" with a *probability*, if some particular quantity is measured, of giving several different outcomes. A relevant illustration in the case at hand would be a gluon with a color state described by:

$(r\bar{b} + b\bar{r})/\sqrt{2}.$

This is read as "red–antiblue plus blue–antired". (The factor of the square root of two is required for normalization, a detail that is not crucial to understand in this discussion.) If one were somehow able to make a direct measurement of the color of a gluon in this state, there would be a 50% chance of it having red-antiblue color charge and a 50% chance of blue-antired color charge.

1.2.2   Color singlet states

It is often said that the stable strongly interacting particles (such as the proton and the neutron, i.e. hadrons) observed in nature are "colorless", but more precisely they are in a "color singlet" state, which is mathematically analogous to a *spin* singlet state.[7] Such states allow interaction with other color singlets, but not with other color states; because long-range gluon interactions do not exist, this illustrates that gluons in the singlet state do not exist either.[7]

The color singlet state is:[7]

$(r\bar{r} + b\bar{b} + g\bar{g})/\sqrt{3}.$

In words, if one could measure the color of the state, there would be equal probabilities of it being red-antired, blue-antiblue, or green-antigreen.

1.2.3   Eight gluon colors

There are eight remaining independent color states, which correspond to the "eight types" or "eight colors" of gluons. Because states can be mixed together as discussed above, there are many ways of presenting these states, which are known as the "color octet". One commonly used list is:[7]

These are equivalent to the Gell-Mann matrices; the translation between the two is that red-antired is the upper-left matrix entry, red-antiblue is the upper middle entry, blue-antigreen is the middle right entry, and so on. The critical feature of these particular eight states is that they are linearly independent, and also independent of the singlet state; there is no way to add any combination of states to produce any other. (It is also impossible to add them to make rr, gg, or bb[8] otherwise the forbidden singlet state could also be made.) There are many other possible choices, but all are mathematically equivalent at least equally complex, and give the same physical results.

1.2.4 Group theory details

Technically, QCD is a gauge theory with SU(3) gauge symmetry. Quarks are introduced as spinor fields in N_f flavors, each in the fundamental representation (triplet, denoted **3**) of the color gauge group, SU(3). The gluons are vector fields in the adjoint representation (octets, denoted **8**) of color SU(3). For a general gauge group, the number of force-carriers (like photons or gluons) is always equal to the dimension of the adjoint representation. For the simple case of SU(N), the dimension of this representation is $N^2 - 1$.

In terms of group theory, the assertion that there are no color singlet gluons is simply the statement that quantum chromodynamics has an SU(3) rather than a U(3) symmetry. There is no known *a priori* reason for one group to be preferred over the other, but as discussed above, the experimental evidence supports SU(3).[7] The U(1) group for electromagnetic field combines with a slightly more complicated group known as SU(2),S stands for "special", which means the corresponding matrices have derterminant 1.

1.3 Confinement

Main article: Color confinement

Since gluons themselves carry color charge, they participate in strong interactions. These gluon-gluon interactions constrain color fields to string-like objects called "flux tubes", which exert constant force when stretched. Due to this force, quarks are confined within composite particles called hadrons. This effectively limits the range of the strong interaction to 1×10^{-15} meters, roughly the size of an atomic nucleus. Beyond a certain distance, the energy of the flux tube binding two quarks increases linearly. At a large enough distance, it becomes energetically more favorable to pull a quark-antiquark pair out of the vacuum rather than increase the length of the flux tube.

Gluons also share this property of being confined within hadrons. One consequence is that gluons are not directly involved in the nuclear forces between hadrons. The force mediators for these are other hadrons called mesons.

Although in the normal phase of QCD single gluons may not travel freely, it is predicted that there exist hadrons that are formed entirely of gluons — called glueballs. There are also conjectures about other exotic hadrons in which real gluons (as opposed to virtual ones found in ordinary hadrons) would be primary constituents. Beyond the normal phase of QCD (at extreme temperatures and pressures), quark–gluon plasma forms. In such a plasma there are no hadrons; quarks and gluons become free particles.

1.4 Experimental observations

Quarks and gluons (colored) manifest themselves by fragmenting into more quarks and gluons, which in turn hadronize into normal (colorless) particles, correlated in jets. As shown in 1978 summer conferences[2] the PLUTO detector at the electron-positron collider DORIS (DESY) produced the first evidence that the hadronic decays of the very narrow resonance $\Upsilon(9.46)$ could be interpreted as three-jet event topologies produced by three gluons. Later published analyses by the same experiment confirmed this interpretation and also the spin 1 nature of the gluon[9][10] (see also the recollection[2] and PLUTO experiments).

In summer 1979 at higher energies at the electron-positron collider PETRA (DESY) again three-jet topologies were observed, now interpreted as qq gluon bremsstrahlung, now clearly visible, by TASSO,[11] MARK-J[12] and PLUTO experiments[13] (later in 1980 also by JADE[14]). The spin 1 of the gluon was confirmed in 1980 by TASSO[15] and PLUTO experiments[16] (see also the review[3]). In 1991 a subsequent experiment at the LEP storage ring at CERN again confirmed this result.[17]

The gluons play an important role in the elementary strong interactions between quarks and gluons, described by QCD and studied particularly at the electron-proton collider HERA at DESY. The number and momentum distribution of the gluons in the proton (gluon density) have been measured by two experiments, H1 and ZEUS,[18] in the years 1996 till today (2012). The gluon contribution to the proton spin has been studied by the HERMES experiment at HERA.[19] The gluon density in the proton (when behaving hadronically) also has been measured.[20]

Color confinement is verified by the failure of free quark searches (searches of fractional charges). Quarks are normally produced in pairs (quark + antiquark) to compensate the quantum color and flavor numbers; however at Fermilab single production of top quarks has been shown (technically this still involves a pair production, but quark and antiquark are of different flavor).[21] No glueball has been demonstrated.

Deconfinement was claimed in 2000 at CERN SPS[22] in heavy-ion collisions, and it implies a new state of matter: quark–gluon plasma, less interacting than in the nucleus, almost as in a liquid. It was found at the Relativistic Heavy Ion Collider (RHIC) at Brookhaven in the years 2004–2010 by four contemporaneous experiments.[23] A quark–gluon plasma state has been confirmed at the CERN Large Hadron Collider (LHC) by the three experiments ALICE, ATLAS and CMS in 2010.[24]

1.5   See also

- Quark

- Hadron

- Meson

- Gauge boson

- Quark model

- Quantum chromodynamics

- Quark–gluon plasma

- Color confinement

- Glueball

- Gluon field

- Gluon field strength tensor

- Exotic hadrons

- Standard Model

- Three-jet events

- Deep inelastic scattering

1.6   References

[1] M. Gell-Mann (1962). "Symmetries of Baryons and Mesons". *Physical Review* **125**(3): 1067–1084. Bibcode:1962PhRv..125. doi:10.1103/PhysRev.125.1067.

[2] B.R. Stella and H.-J. Meyer (2011). "ϒ(9.46 GeV) and the gluon discovery (a critical recollection of PLUTO results)". *European Physical Journal H* **36** (2): 203–243. arXiv:1008.1869v3. Bibcode:2011EPJH...36..203S. doi:10.1140/epjh/e2011-10029-3.

[3] P. Söding (2010). "On the discovery of the gluon". *European Physical Journal H* **35** (1): 3–28. Bibcode:2010EPJH...35....3S. doi:10.1140/epjh/e2010-00002-5.

[4] W.-M. Yao; et al. (2006). "Review of Particle Physics" (PDF). *Journal of Physics G* **33**: 1. arXiv:astro-ph/0601168. Bibcode:2 doi:10.1088/0954-3899/33/1/001.

[5] F. Yndurain (1995). "Limits on the mass of the gluon". *Physics Letters B* **345** (4): 524. Bibcode:1995PhLB..345..524Y. doi:10.1016/0370-2693(94)01677-5.

[6] C.R. Nave. "The Color Force". *HyperPhysics*. Georgia State University, Department of Physics. Retrieved 2012-04-02.

[7] David Griffiths (1987). *Introduction to Elementary Particles*. John Wiley & Sons. pp. 280–281. ISBN 0-471-60386-4.

[8] J. Baez. "Why are there eight gluons and not nine?". Retrieved 2009-09-13.

[9] Ch. Berger *et al.* (PLUTO Collaboration) (1979). "Jet analysis of the $\Upsilon(9.46)$ decay into charged hadrons".*Physics Letters* **82** (3–4): 449. Bibcode:1979PhLB...82..449B. doi:10.1016/0370-2693(79)90265-X.

[10] Ch. Berger *et al.* (PLUTO Collaboration) (1981). "Topology of the Υ decay".*Zeitschrift für Physik C***8**(2): 101.Bibcode:1981 doi:10.1007/BF01547873.

[11] R. Brandelik *et al.* (TASSO collaboration) (1979). "Evidence for Planar Events in e^+e^- Annihilation at High Energies". *Physics Letters B* **86** (2): 243–249. Bibcode:1979PhLB...86..243B. doi:10.1016/0370-2693(79)90830-X.

[12] D.P. Barber *et al.* (MARK-J collaboration) (1979). "Discovery of Three-Jet Events and a Test of Quantum Chromodynamics at PETRA". *Physical Review Letters* **43** (12): 830. Bibcode:1979PhRvL..43..830B. doi:10.1103/PhysRevLett.43.830.

[13] Ch. Berger *et al.* (PLUTO Collaboration) (1979). "Evidence for Gluon Bremsstrahlung in e^+e^- Annihilations at High Energies". *Physics Letters B* **86** (3–4): 418. Bibcode:1979PhLB...86..418B. doi:10.1016/0370-2693(79)90869-4.

[14] W. Bartel *et al.* (JADE Collaboration) (1980). "Observation of planar three-jet events in e^+e^- annihilation and evidence for gluon bremsstrahlung". *Physics Letters B* **91**: 142. Bibcode:1980PhLB...91..142B. doi:10.1016/0370-2693(80)90680-2.

[15] R. Brandelik *et al.* (TASSO Collaboration) (1980). "Evidence for a spin-1 gluon in three-jet events". *Physics Letters B* **97** (3–4): 453. Bibcode:1980PhLB...97..453B. doi:10.1016/0370-2693(80)90639-5.

[16] Ch. Berger *et al.* (PLUTO Collaboration) (1980). "A study of multi-jet events in e^+e^- annihilation". *Physics Letters B* **97** (3–4): 459. Bibcode:1980PhLB...97..459B. doi:10.1016/0370-2693(80)90640-1.

[17] G. Alexander *et al.* (OPAL Collaboration) (1991). "Measurement of Three-Jet Distributions Sensitive to the Gluon Spin in e^+e^- Annihilations at $\sqrt{s} = 91$ GeV". *Zeitschrift für Physik C* **52** (4): 543. Bibcode:1991ZPhyC..52..543A. doi:10.1007/BF01562326.

[18] L. Lindeman (H1 and ZEUS collaborations) (1997). "Proton structure functions and gluon density at HERA". *Nuclear Physics B Proceedings Supplements* **64**: 179–183. Bibcode:1998NuPhS..64..179L. doi:10.1016/S0920-5632(97)01057-8.

[19] http://www-hermes.desy.de

[20] C. Adloff *et al.* (H1 collaboration) (1999). "Charged particle cross sections in the photoproduction and extraction of the gluon density in the photon". *European Physical Journal C* **10**: 363–372. arXiv:hep-ex/9810020. Bibcode:1999EPJC...10..363H. doi:10.1007/s100520050761.

[21] M. Chalmers (6 March 2009). "Top result for Tevatron". *Physics World*. Retrieved 2012-04-02.

[22] M.C. Abreu; et al. (2000). "Evidence for deconfinement of quark and antiquark from the J/Ψ suppression pattern measured in Pb-Pb collisions at the CERN SpS". *Physics Letters B* **477**: 28–36. Bibcode:2000PhLB..477...28A. doi:10.1016/S0370-2693(00)00237-9.

[23] D. Overbye (15 February 2010). "In Brookhaven Collider, Scientists Briefly Break a Law of Nature". *New York Times*. Retrieved 2012-04-02.

[24] "LHC experiments bring new insight into primordial universe" (Press release). CERN. 26 November 2010. Retrieved 2012-04-02.

1.7 Further reading

- A. Ali and G. Kramer (2011). "JETS and QCD: A historical review of the discovery of the quark and gluon jets and its impact on QCD". *European Physical Journal H* **36** (2): 245–326. arXiv:1012.2288. Bibcode:2011EPJH...36..2 doi:10.1140/epjh/e2011-10047-1.

Chapter 2

Gluon field

Further information: Ricci calculus, Special unitary group and Quantum chromodynamics

In theoretical particle physics, the **gluon field** is a four vector field characterizing the propagation of gluons in the strong interaction between quarks. It plays the same role in quantum chromodynamics as the electromagnetic four-potential in quantum electrodynamics - the gluon field constructs the gluon field strength tensor.

Throughout, Latin indices take values 1, 2, ..., 8 for the eight gluon color charges, while Greek indices take values 0 for timelike components and 1, 2, 3 for spacelike components of four-dimensional vectors and tensors in spacetime. Throughout all equations, the summation convention is used on all color and tensor indices, unless explicitly stated otherwise.

2.1 Introduction

Gluons can have eight colour charges so there are eight fields, in contrast to photons which are neutral and so there is only one photon field.

The gluon fields for each color charge each have a "timelike" component analogous to the electric potential, and three "spacelike" components analogous to the magnetic vector potential. Using similar symbols:[1]

$$\boldsymbol{\mathcal{A}}^n(\mathbf{r}, t) = [\underbrace{\mathcal{A}_0^n(\mathbf{r}, t)}_{\text{timelike}}, \underbrace{\mathcal{A}_1^n(\mathbf{r}, t), \mathcal{A}_2^n(\mathbf{r}, t), \mathcal{A}_3^n(\mathbf{r}, t)}_{\text{spacelike}}] = [\phi^n(\mathbf{r}, t), \mathbf{A}^n(\mathbf{r}, t)]$$

where $n = 1, 2, \ldots 8$ are not exponents but enumerate the eight gluon color charges, and all components depend on the position vector $\mathbf{r}$ of the gluon and time t. Each $\mathcal{A}_\alpha^a$ is a scalar field, for some component of spacetime and gluon color charge.

The Gell-Mann matrices λ^a are eight 3×3 matrices which form matrix representations of the $SU(3)$ group. They are also generators of the SU(3) group, in the context of quantum mechanics and field theory; a generator can be viewed as an operator corresponding to a symmetry transformation (see symmetry in quantum mechanics). These matrices play an important role in QCD as QCD is a gauge theory of the SU(3) gauge group obtained by taking the color charge to define a local symmetry: each Gell-Mann matrix corresponds to a particular gluon color charge, which in turn can be used to define color charge operators. Generators of a group can also form a basis for a vector space, so the overall gluon field is a "superposition" of all the color fields. In terms of the Gell-Mann matrices (divided by 2 for convenience),

$$t_a = \frac{\lambda_a}{2},$$

the components of the gluon field are represented by 3×3 matrices, given by:

$$\mathcal{A}_\alpha = t_a \mathcal{A}_\alpha^a \equiv t_1 \mathcal{A}_\alpha^1 + t_2 \mathcal{A}_\alpha^2 + \cdots t_8 \mathcal{A}_\alpha^8$$

or collecting these into a vector of four 3×3 matrices:

$$\boldsymbol{\mathcal{A}}(\mathbf{r}, t) = [\mathcal{A}_0(\mathbf{r}, t), \mathcal{A}_1(\mathbf{r}, t), \mathcal{A}_2(\mathbf{r}, t), \mathcal{A}_3(\mathbf{r}, t)]$$

the gluon field is:

$$\boldsymbol{\mathcal{A}} = t_a \boldsymbol{\mathcal{A}}^a .$$

2.2 Gauge covariant derivative in QCD

Below the definitions (and most of the notation) follow K. Yagi, T. Hatsuda, Y. Miake[2] and Greiner, Schäfer.[3]

The gauge covariant derivative $D\mu$ is required to transform quark fields in manifest covariance; the partial derivatives that form the four-gradient $\partial\mu$ alone are not enough. The components which act on the color triplet quark fields are given by:

$$D_\mu = \partial_\mu \pm i g_s t_a \mathcal{A}_\mu^a ,$$

wherein i is the imaginary unit, and

$$g_s = \sqrt{4\pi\alpha_s}$$

is the dimensionless coupling constant for QCD. Different authors choose different signs. The partial derivative term includes a 3×3 identity matrix, conventionally not written for simplicity.

The quark fields in triplet representation are written as column vectors:

$$\psi = \begin{pmatrix} \psi_1 \\ \psi_2 \\ \psi_3 \end{pmatrix}, \overline{\psi} = \begin{pmatrix} \overline{\psi}_1^* \\ \overline{\psi}_2^* \\ \overline{\psi}_3^* \end{pmatrix}$$

The quark field ψ belongs to the fundamental representation (**3**) and the antiquark field ψ belongs to the complex conjugate representation (**3***), complex conjugate is denoted by * (not overbar).

2.3 Gauge transformations

Main article: gauge theory

The gauge transformation of each gluon field $\mathcal{A}_\alpha^n$ which leaves the gluon field strength tensor unchanged is;[3]

$$\mathcal{A}_\alpha^n \rightarrow e^{i\bar{\theta}(\mathbf{r},t)} \left(\mathcal{A}_\alpha^n + \frac{i}{g_s} \partial_\alpha \right) e^{-i\bar{\theta}(\mathbf{r},t)}$$

where

$$\bar{\theta}(\mathbf{r}, t) = t_n \theta^n(\mathbf{r}, t),$$

is a 3×3 matrix constructed from the t^n matrices above and $\theta^n = \theta^n(\mathbf{r}, t)$ are eight gauge functions dependent on spatial position $\mathbf{r}$ and time t. Matrix exponentiation is used in the transformation. The gauge covariant derivative transforms similarly. The functions θ^n here are similar to the gauge function $\chi(\mathbf{r}, t)$ when changing the electromagnetic four potential A, in spacetime components:

$$A'_\alpha(\mathbf{r}, t) = A_\alpha(\mathbf{r}, t) - \partial_\alpha \chi(\mathbf{r}, t)$$

leaving the electromagnetic tensor F invariant.

The quark fields are invariant under the gauge transformation;[3]

$$\psi(\mathbf{r}, t) \to e^{ig\bar{\theta}(\mathbf{r},t)} \psi(\mathbf{r}, t)$$

2.4   See also

- Quark confinement

- Gell-Mann matrices

- Field (physics)

- Einstein tensor

- Symmetry in quantum mechanics

- Wilson loop

- Wess–Zumino gauge

2.5   References

2.5.1   Notes

[1] B.R. Martin, G. Shaw (2009). *Particle Physics.* Manchester Physics Series (3rd ed.). John Wiley & Sons. pp. 380–384. ISBN 978-0-470-03294-7.

[2] K. Yagi, T. Hatsuda, Y. Miake (2005). *Quark-Gluon Plasma: From Big Bang to Little Bang.* Cambridge monographs on particle physics, nuclear physics, and cosmology **23**. Cambridge University Press. pp. 17–18. ISBN 0-521-561-086.

[3] W. Greiner, G. Schäfer (1994). "4". *Quantum Chromodynamics.* Springer. ISBN 3-540-57103-5.

2.5.2   Further reading

Books

- W. N. Cottingham, D. A. Greenwood (2007). *An Introduction to the Standard Model of Particle Physics.* Cambridge University Press. ISBN 113-946-221-0.

- H. Fritzsch (1982). *Quarks: the stuff of matter*. Allen lane. ISBN 0-7139-15331.

- S. Sarkar, H. Satz, B. Sinha (2009). *The Physics of the Quark-Gluon Plasma: Introductory Lectures*. Springer. ISBN 3642022855.

- J. Thanh Van Tran (editor) (1987). *Hadrons, Quarks and Gluons: Proceedings of the Hadronic Session of the Twenty-Second Rencontre de Moriond, Les Arcs-Savoie-France*. Atlantica Séguier Frontières. ISBN 2863320483.

- R. Alkofer, H. Reinhart (1995). *Chiral Quark Dynamics*. Springer. ISBN 3540601376.

- K. Chung (2008). *Hadronic Production of $\psi(2S)$ Cross Section and Polarization*. ProQuest. ISBN 0549597743.

- J. Collins (2011). *Foundations of Perturbative QCD*. Cambridge University Press. ISBN 0521855330.

- W.N.A. Cottingham, D.A.A. Greenwood (1998). *Standard Model of Particle Physics*. Cambridge University Press. ISBN 0521588324.

Selected papers

- J.P. Maa, Q. Wang, G.P. Zhang (2012). "QCD evolutions of twist-3 chirality-odd operators". *Physics Letters B*. arXiv:1210.1006. Bibcode:2013PhLB..718.1358M. doi:10.1016/j.physletb.2012.12.007.

- M. D'Elia, A. Di Giacomo, E. Meggiolaro (1997). "Field strength correlators in full QCD". *Physics Letters B*. arXiv:hep-lat/9705032. Bibcode:1997PhLB..408..315D. doi:10.1016/S0370-2693(97)00814-9.

- A. Di Giacomo, M. D'elia, H. Panagopoulos, E. Meggiolaro (1998). "Gauge Invariant Field Strength Correlators In QCD". arXiv:hep-lat/9808056.

- M. Neubert (1993). "A Virial Theorem for the Kinetic Energy of a Heavy Quark inside Hadrons". *Physics Letters B*. arXiv:hep-ph/9311232.

- M. Neubert, N. Brambilla, H.G. Dosch, A. Vairo (1998). "Field strength correlators and dual effective dynamics in QCD".*Physical Review D*.arXiv:hep-ph/9311232.Bibcode:1998PhRvD..58c4010B.doi:10.1103/PhysRevD.58.0

- V. Dzhunushaliev (2011). "Gluon field distribution between three infinitely spaced quarks". arXiv:1101.5845.

2.6 External links

- K. Ellis (2005). "QCD" (PDF). Fermilab.

- "Chapter 2: The QCD Lagrangian" (PDF). Technische Universität München. Retrieved 2013-10-17.

Chapter 3

Gluon field strength tensor

Further information: Ricci calculus, Special unitary group and Quantum chromodynamics

In theoretical particle physics, the **gluon field strength tensor** is a second order tensor field characterizing the gluon interaction between quarks.

The strong interaction is one of the fundamental interactions of nature, and the quantum field theory (QFT) to describe it is called *quantum chromodynamics* (QCD). Quarks interact with each other by the strong force due to their color charge, mediated by gluons. Gluons themselves possess color charge and can mutually interact.

The gluon field strength tensor is a rank 2 tensor field on the spacetime with values in the adjoint bundle of the chromodynamical SU(3) gauge group (see vector bundle for necessary definitions). Throughout, Latin indices (typically a, b, c, n) take values 1, 2, ..., 8 for the eight gluon color charges, while Greek indices (typically α, β, μ, ν) take values 0 for timelike components and 1, 2, 3 for spacelike components of four-vectors and four-dimensional spacetime tensors. Throughout all equations, the summation convention is used on all color and tensor indices, unless explicitly stated there is no sum to be taken.

3.1 Definition

Below the definitions (and most of the notation) follow K. Yagi, T. Hatsuda, Y. Miake[1] and Greiner, Schäfer.[2]

3.1.1 Tensor components

The tensor is denoted G, (or F, $\mathcal{F}$, or some variant), and has components defined proportional to the commutator of the quark covariant derivative D_μ:[2][3]

$$G_{\alpha\beta} = \pm \frac{1}{g_s}[D_\alpha, D_\beta] \,,$$

where:

$$D_\mu = \partial_\mu \pm i g_s t_a \mathcal{A}_\mu^a \,,$$

in which

- i is the imaginary unit;

- *gs* is the coupling constant of the strong force;

- *ta* = λ*a*/2 are the Gell-Mann matrices λ*a* divided by 2;

- *a* is a color index in the adjoint representation of SU(3) which take values 1, 2, ..., 8 for the eight generators of the group, namely the Gell-Mann matrices;

- μ is a spacetime index, 0 for timelike components and 1, 2, 3 for spacelike components;

- $\mathcal{A}_\mu = t_a \mathcal{A}_\mu^a$ expresses the gluon field, a spin-1 gauge field or, in differentially-geometric parlance, a connection in the SU(3) principal bundle;

- $\mathcal{A}_\mu$ are its four (coordinate-system dependent) components, that in a fixed gauge are 3×3 traceless Hermitian matrix-valued functions, while $\mathcal{A}_\mu^a$ are 32 real-valued functions, the four components for each of the eight four-vector fields.

Different authors choose different signs.

Expanding the commutator gives;

$$G_{\alpha\beta} = \partial_\alpha \mathcal{A}_\beta - \partial_\beta \mathcal{A}_\alpha \pm ig_s [\mathcal{A}_\alpha, \mathcal{A}_\beta]$$

Substituting $t_a \mathcal{A}_\alpha^a = \mathcal{A}_\alpha$ and using the commutation relation $[t_a, t_b] = i f^{abc} t_c$ for the Gell-Mann matrices (with a relabeling of indices), in which f^{abc} are the structure constants of SU(3), each of the gluon field strength components can be expressed as a linear combination of the Gell-Mann matrices as follows:

$$\begin{aligned} G_{\alpha\beta} &= \partial_\alpha t_a \mathcal{A}_\beta^a - \partial_\beta t_a \mathcal{A}_\alpha^a \pm ig_s [t_b, t_c] \mathcal{A}_\alpha^b \mathcal{A}_\beta^c \\ &= t_a \left(\partial_\alpha \mathcal{A}_\beta^a - \partial_\beta \mathcal{A}_\alpha^a \pm i^2 g_s \mathcal{A}_\alpha^b \mathcal{A}_\beta^c \right) \quad , \\ &= t_a G_{\alpha\beta}^a \end{aligned}$$

so that:[4][5]

$$G_{\alpha\beta}^a = \partial_\alpha \mathcal{A}_\beta^a - \partial_\beta \mathcal{A}_\alpha^a \mp g_s f^{abc} \mathcal{A}_\alpha^b \mathcal{A}_\beta^c ,$$

where again *a, b, c* = 1, 2, ..., 8 are color indices. As with the gluon field, in a specific coordinate system and fixed gauge $G\alpha\beta$ are 3×3 traceless Hermitian matrix-valued functions, while $G^a\alpha\beta$ are real-valued functions, the components of eight four-dimensional second order tensor fields.

3.1.2 Differential forms

The gluon color field can be described using the language of differential forms, specifically as an adjoint bundle-valued curvature 2-form (note that fibers of the adjoint bundle are the **su**(3) Lie algebra);

$$\mathbf{G} = \mathrm{d}\boldsymbol{\mathcal{A}} \mp g_s \, \boldsymbol{\mathcal{A}} \wedge \boldsymbol{\mathcal{A}} ,$$

where $\boldsymbol{\mathcal{A}}$ is the gluon field, a vector potential 1-form corresponding to **G** and $\wedge$ is the (antisymmetric) wedge product of this algebra, producing the structure constants f^{abc}. The Cartan-derivative of the field form (i.e. essentially the divergence of the field) would be zero in the absence of the "gluon terms", i.e. those $\boldsymbol{\mathcal{A}}$ which represent the non-abelian character of the SU(3).

3.1.3   Comparison with the electromagnetic tensor

This almost parallels the electromagnetic field tensor (also denoted F) in quantum electrodynamics, given by the electromagnetic four-potential A describing a spin-1 photon;

$$F_{\alpha\beta} = \partial_\alpha A_\beta - \partial_\beta A_\alpha \,,$$

or in the language of differential forms:

$$\mathbf{F} = \mathrm{d}\mathbf{A} \,.$$

The key difference between quantum electrodynamics and quantum chromodynamics is that the gluon field strength has extra terms which lead to self-interactions between the gluons and asymptotic freedom. This is a complication of the strong force making it inherently non-linear, contrary to the linear theory of the electromagnetic force. QCD is a non-abelian gauge theory. The word *non-abelian* in group-theoretical language means that the group operation in not commutative, that makes the corresponding Lie algebra non-trivial.

3.2   QCD Lagrangian density

See also: classical field theory

Characteristic of field theories, the dynamics of the field strength are summarized by a suitable Lagrangian density and substitution into the Euler–Lagrange equation (for fields) obtains the equation of motion for the field. The Lagrangian density for massless quarks, bound by gluons, is:[2]

$$\mathcal{L} = -\frac{1}{2}\mathrm{tr}\left(G_{\alpha\beta}G^{\alpha\beta}\right) + \bar{\psi}\left(iD_\mu\right)\gamma^\mu\psi$$

where "tr" denotes trace of the 3×3 matrix $G\alpha\beta G^{\alpha\beta}$, and γ^μ are the 4×4 gamma matrices.

3.3   Gauge transformations

Main article: Gauge theory

In contrast to QED, the gluon field strength tensor is not gauge invariant by itself. Only the product of two contracted over all indices is gauge invariant.

3.4   Equations of motion

The equations[1] governing the evolution of the quark fields are:

$$(i\hbar\gamma^\mu D_\mu - mc)\psi = 0$$

which is like the Dirac equation, and the equations for the gluon field strength tensor are:

$$[D_\mu, G^{\mu\nu}] = g_s j^\nu$$

which are similar to the Maxwell equations (when written in tensor notation), more specifically the Yang–Mills equations for quarks and gluons. The color charge four-current is the source of the gluon field strength tensor, analogous to the electromagnetic four-current as the source of the electromagnetic tensor, given by:

$$j^\nu = t^b j_b^\nu\,, \quad j_b^\nu = \bar{\psi}\gamma^\nu t^b \psi\,,$$

which is a conserved current since color charge is conserved, in other words the color four-current must satisfy the continuity equation:

$$\partial_\nu j^\nu = 0\,.$$

3.5 See also

- Quark confinement

- Gell-Mann matrices

- Field (physics)

- Yang–Mills field

- Eightfold Way (physics)

- Einstein tensor

- Wilson loop

- Wess–Zumino gauge

3.6 References

3.6.1 Notes

[1] K. Yagi, T. Hatsuda, Y. Miake (2005). *Quark-Gluon Plasma: From Big Bang to Little Bang*. Cambridge monographs on particle physics, nuclear physics, and cosmology **23**. Cambridge University Press. pp. 17–18. ISBN 0-521-561-086.

[2] W. Greiner, G. Schäfer (1994). "4". *Quantum Chromodynamics*. Springer. ISBN 3-540-57103-5.

[3] S.O. Bilson-Thompson, D.B. Leinweber, A.G. Williams (2003). "Highly improved lattice field-strength tensor". *Annals of Physics*. 304(1) (Adelaide, Australia: Elsevier). pp. 1–21.

[4] M. Eidemüller, H.G. Dosch, M. Jamin (1999). "The field strength correlator from QCD sum rules". *Nucl.Phys.Proc.Suppl.86:421-425,2000* (Heidelberg, Germany). arXiv:hep-ph/9908318.

[5] M. Shifman (2012). *Advanced Topics in Quantum Field Theory: A Lecture Course*. Cambridge University Press. ISBN 0521190843.

3.6.2   Further reading

Books

- H. Fritzsch (1982). *Quarks: the stuff of matter*. Allen lane. ISBN 0-7139-15331.

- B.R. Martin, G. Shaw (2009). *Particle Physics*. Manchester Physics Series (3rd ed.). John Wiley & Sons. ISBN 978-0-470-03294-7.

- S. Sarkar, H. Satz, B. Sinha (2009). *The Physics of the Quark-Gluon Plasma: Introductory Lectures*. Springer. ISBN 3642022855.

- J. Thanh Van Tran (editor) (1987). *Hadrons, Quarks and Gluons: Proceedings of the Hadronic Session of the Twenty-Second Rencontre de Moriond, Les Arcs-Savoie-France*. Atlantica Séguier Frontières. ISBN 2863320483.

- R. Alkofer, H. Reinhart (1995). *Chiral Quark Dynamics*. Springer. ISBN 3540601376.

- K. Chung (2008). *Hadronic Production of $\psi(2S)$ Cross Section and Polarization*. ProQuest. ISBN 0549597743.

- J. Collins (2011). *Foundations of Perturbative QCD*. Cambridge University Press. ISBN 0521855330.

- W.N.A. Cottingham, D.A.A. Greenwood (1998). *Standard Model of Particle Physics*. Cambridge University Press. ISBN 0521588324.

Selected papers

- J.P. Maa, Q. Wang, G.P. Zhang (2012). "QCD evolutions of twist-3 chirality-odd operators". *Physics Letters B*. arXiv:1210.1006. Bibcode:2013PhLB..718.1358M. doi:10.1016/j.physletb.2012.12.007.

- M. D'Elia, A. Di Giacomo, E. Meggiolaro (1997). "Field strength correlators in full QCD". *Physics Letters B*. arXiv:hep-lat/9705032. Bibcode:1997PhLB..408..315D. doi:10.1016/S0370-2693(97)00814-9.

- A. Di Giacomo, M. D'elia, H. Panagopoulos, E. Meggiolaro (1998). "Gauge Invariant Field Strength Correlators In QCD". arXiv:hep-lat/9808056.

- M. Neubert (1993). "A Virial Theorem for the Kinetic Energy of a Heavy Quark inside Hadrons". *Physics Letters B*. arXiv:hep-ph/9311232. Bibcode:1994PhLB..322..419N. doi:10.1016/0370-2693(94)91174-6.

- M. Neubert, N. Brambilla, H.G. Dosch, A. Vairo (1998). "Field strength correlators and dual effective dynamics in QCD".*Physical Review D*.arXiv:hep-ph/9311232.Bibcode:1998PhRvD..58c4010B.doi:10.1103/PhysRevD.58.0

- M. Neubert (1996). "QCD sum-rule calculation of the kinetic energy and chromo-interaction of heavy quarks inside mesons" (PDF). *Physics Letters B*.

3.7   External links

- K. Ellis (2005). "QCD" (PDF). Fermilab.

- "Chapter 2: The QCD Lagrangian" (PDF). Technische Universität München. Retrieved 2013-10-17.

Chapter 4

Color confinement

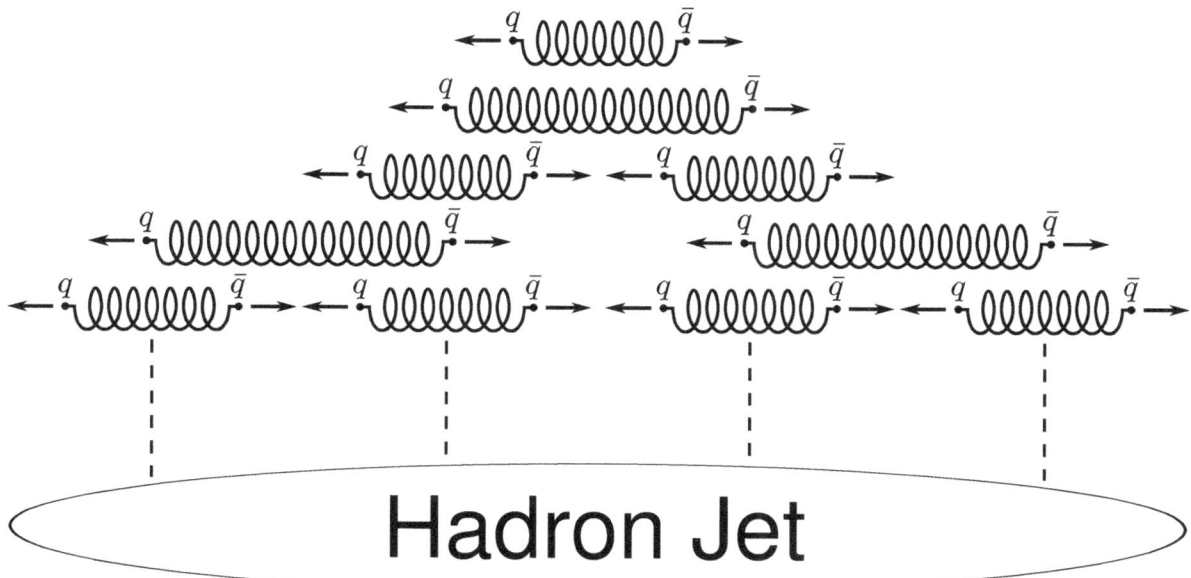

The color force favors confinement because at a certain range it is more energetically favorable to create a quark-antiquark pair than to continue to elongate the color flux tube. This is analoguous to the behavior of an elongated rubber-band.

Color confinement, often simply called **confinement**, is the phenomenon that color charged particles (such as quarks) cannot be isolated singularly, and therefore cannot be directly observed.[1] Quarks, by default, clump together to form groups, or hadrons. The two types of hadrons are the mesons (one quark, one antiquark) and the baryons (three quarks).

The constituent quarks in a group cannot be separated from their parent hadron, and this is why quarks currently cannot be studied or observed in any more direct way than at a hadron level.[2]

4.1 Origin

The reasons for quark confinement are somewhat complicated; no analytic proof exists that quantum chromodynamics should be confining. The current theory is that confinement is due to the force-carrying gluons having color charge. As any two electrically charged particles separate, the electric fields between them diminish quickly, allowing (for example) electrons to become unbound from atomic nuclei. However, as a quark-antiquark pair separates, the gluon field forms a narrow tube (or string) of color field between them. This is quite different from the behavior of the electric field of

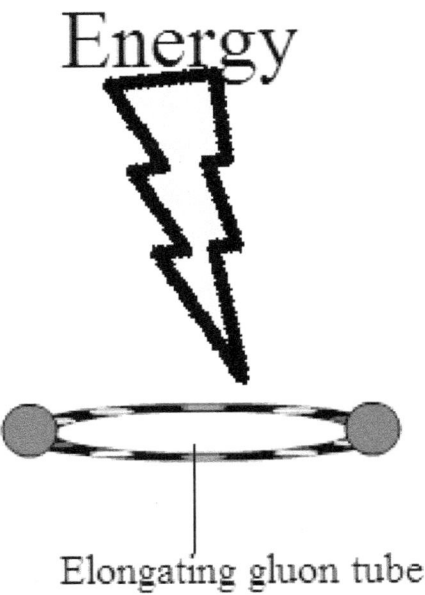

An animation of color confinement. Energy is supplied to the quarks, and the gluon tube elongates until it reaches a point where it "snaps" and forms a quark-antiquark pair.

a pair of positive and negative electric charges, which extends into the whole surrounding space and diminishes at large distances. Because of this behavior of the gluon field, a strong force between the quark pair acts constantly—regardless of their distance[3][4]—with a force of around 10,000 newtons. [5]

When two quarks become separated, as happens in particle accelerator collisions, at some point it is more energetically favorable for a new quark–antiquark pair to spontaneously appear, than to allow the tube to extend further. As a result of this, when quarks are produced in particle accelerators, instead of seeing the individual quarks in detectors, scientists see "jets" of many color-neutral particles (mesons and baryons), clustered together. This process is called *hadronization*, *fragmentation*, or *string breaking*, and is one of the least understood processes in particle physics.

The confining phase is usually defined by the behavior of the action of the Wilson loop, which is simply the path in spacetime traced out by a quark–antiquark pair created at one point and annihilated at another point. In a non-confining theory, the action of such a loop is proportional to its perimeter. However, in a confining theory, the action of the loop is instead proportional to its area. Since the area will be proportional to the separation of the quark–antiquark pair, free quarks are suppressed. Mesons are allowed in such a picture, since a loop containing another loop in the opposite direction will have only a small area between the two loops.

4.2   Models exhibiting confinement

Besides QCD in four spacetime dimensions, another model which exhibits confinement is the Schwinger model.[6] Compact Abelian gauge theories also exhibit confinement in 2 and 3 spacetime dimensions.[7] Confinement has recently been found in elementary excitations of magnetic systems called spinons.[8]

4.3 Models of fully screened quarks

Besides the quark confinement idea, there is a potential possibility, that color charge of quarks gets fully screened by the gluonic color, surrounding the quark. Exact solutions of SU(3) classical Yang–Mills theory, which provide full screening (by gluon fields) of the color charge of a quark have been found.[9] However, such classical solutions do not take into account non-trivial properties of QCD vacuum. Therefore, a significance of such full gluonic screening solutions for a separated quark is not clear.

4.4 See also

- Gluon field strength tensor
- Asymptotic freedom
- Center vortices
- Deconfining phase
- Quantum mechanics
- Particle physics
- Fundamental force
- Dual superconducting model
- Beta-function
- Infrared safety

4.5 References

[1] V. Barger, R. Phillips (1997). *Collider Physics.* Addison–Wesley. ISBN 0-201-14945-1.

[2] T.-Y. Wu, W.-Y. Pauchy Hwang (1991). *Relativistic quantum mechanics and quantum fields.* World Scientific. p. 321. ISBN 981-02-0608-9.

[3] T. Muta (2009). *Foundations of quantum chromodynamics: an introduction to perturbative methods in gauge theories* (3rd ed.). World Scientific. ISBN 978-981-279-353-9.

[4] A. Smilga (2001). *Lectures on quantum chromodynamics.* World Scientific. ISBN 978-981-02-4331-9.

[5] Fritzsch, op. cite, p. 164. The author states that the force between differently coloured quarks remains constant at any distance after they travel only a tiny distance from each other, and is equal to that need to raise one ton, which is 1000 kg x 9.8 m/s^2 = ~10,000 N.

[6] Wilson, Kenneth G. (1974-10-15). "Confinement of Quarks". *Physical Review D* (College Park, MD, USA: American Physical Society) **10**: 2445–2459. Bibcode:1974PhRvD..10.2445W. doi:10.1103/PhysRevD.10.2445. ISSN 1550-2368. OCLC 55589778. Retrieved 2014-04-12.

[7] Schön, Verena; Michael, Thies (2000-08-22). "2d Model Field Theories at Finite Temperature and Density (Section 2.5)". arXiv:hep-th/0008175v1 [hep-th].

[8] Lake, Bella; Tsvelik, Alexei M.; Notbohm, Susanne; Tennant, D. Alan; Perring, Toby G.; Reehuis, Manfred; Sekar, Chinnathambi; Krabbes, Gernot; Büchner, Bernd (2009-11-29). "Confinement of fractional quantum number particles in a condensed-matter system". *Nature Physics* (London, UK: Nature Publishing Group) **6** (1): 50–55. arXiv:0908.1038. Bibcode:2010NatPh...6...50L. doi:10.1038/nphys1462. ISSN 1745-2481. OCLC 150143123. Retrieved 2014-04-12. (subscription required (help)).

[9] Cahill, Kevin (1978-08-28). "Example of Color Screening". *Physical Review Letters* (American Physical Society) **41** (9): 599–601. Bibcode:1978PhRvL..41..599C. doi:10.1103/PhysRevLett.41.599. ISSN 1079-7114. OCLC 31492939. Retrieved 2014-04-12. (subscription required (help)).

4.6   External links

- Quarks

Chapter 5

Quark–gluon plasma

A **quark–gluon plasma (QGP)** or **quark soup**[1] is a state of matter in quantum chromodynamics (QCD) which is hypothesized to exist at extremely high temperature, density, or both temperature and density. This state is thought to consist of asymptotically free quarks and gluons, which are several of the basic building blocks of matter. It is believed that up to a few milliseconds after the Big Bang, known as the Quark epoch, the Universe was in a quark–gluon plasma state. On June, 2015 an international team of physicists have produced quark-gluon plasma at the Large Hadron Collider by colliding protons with lead nuclei at high energy inside the supercollider's Compact Muon Solenoid detector. They also discovered that this new state of matter behaves like a fluid.[2]

The strength of the color force means that unlike the gas-like plasma, quark–gluon plasma behaves as a near-ideal Fermi liquid, although research on flow characteristics is ongoing.[3] In the quark matter phase diagram, QGP is placed in the high-temperature, high-density regime; whereas, ordinary matter is a cold and rarefied mixture of nuclei and vacuum, and the hypothetical quark stars would consist of relatively cold, but dense quark matter.

Experiments at CERN's Super Proton Synchrotron (SPS) first tried to create the QGP in the 1980s and 1990s: the results led CERN to announce indirect evidence for a "new state of matter"[4] in 2000. Current experiments (2011) at the Brookhaven National Laboratory's Relativistic Heavy Ion Collider (RHIC) on Long Island (NY, USA) and at CERN's recent Large Hadron Collider near Geneva (Switzerland) are continuing this effort,[5][6] by colliding relativistically accelerated gold (at RHIC) or lead (at LHC) with each other or with protons. Although the results have yet to be independently verified as of February 2010, scientists at Brookhaven RHIC have tentatively claimed to have created a quark–gluon plasma with an approximate temperature of 4 trillion (4×10^{12}) degrees Kelvin.[6]

As already mentioned, three new experiments running on CERN's Large Hadron Collider (LHC), on the spectrometers ALICE,[7] ATLAS and CMS, will continue studying properties of QGP. Starting in November 2010, CERN temporarily ceased colliding protons, and began colliding lead Ions for the ALICE experiment. They were looking to create a QGP and were expected to stop December 6, colliding protons again in January.[8] A new record breaking temperature was set by ALICE: A Large Ion Collider Experiment at CERN on August, 2012 in the ranges of 5.5 trillion (5.5×10^{12}) degrees Kelvin as claimed in their Nature PR.[9]

5.1 General introduction

Quark–gluon plasma is a state of matter in which the elementary particles that make up the hadrons of baryonic matter are freed of their strong attraction for one another under extremely high energy densities. These particles are the quarks and gluons that compose baryonic matter.[10] In normal matter quarks are *confined*; in the QGP quarks are *deconfined*. In classical QCD quarks are the Fermionic components of mesons and baryons while the gluons are considered the Bosonic components of such particles. The gluons are the force carriers, or bosons, of the QCD color force, while the quarks by themselves are their Fermionic matter counterparts.

Although the experimental high temperatures and densities predicted as producing a quark–gluon plasma have been realized in the laboratory, the resulting matter does *not* behave as a quasi-ideal state of free quarks and gluons, but, rather, as an almost perfect dense fluid.[11] Actually, the fact that the quark–gluon plasma will not yet be "free" at temperatures realized at present accelerators was predicted in 1984 as a consequence of the remnant effects of confinement.[12][13]

5.1.1   Relation to normal plasma

A plasma is matter in which charges are screened due to the presence of other mobile charges; for example: Coulomb's Law is suppressed by the screening to yield a distance-dependent charge ($Q \rightarrow Q \times \exp(-r/\alpha)$, i.e, the charge Q is reduced exponentially with the distance divided by a screening length α). In a QGP, the color charge of the quarks and gluons is screened. The QGP has other analogies with a normal plasma. There are also dissimilarities because the color charge is non-abelian, whereas the electric charge is abelian. Outside a finite volume of QGP the color-electric field is not screened, so that a volume of QGP must still be color-neutral. It will therefore, like a nucleus, have integer electric charge.

5.1.2   Theory

One consequence of this difference is that the color charge is too large for perturbative computations which are the mainstay of QED. As a result, the main theoretical tools to explore the theory of the QGP is lattice gauge theory.[14] The transition temperature (approximately 175 MeV) was first predicted by lattice gauge theory. Since then lattice gauge theory has been used to predict many other properties of this kind of matter. The AdS/CFT correspondence conjecture may provide insights in QGP, morever the ultimate goal of the fluid/gravity correspondence is to understand QGP. The QCP is believed to be a phase of QCD which is completely locally thermalized and thus suitable for an effective fluid dynamic description.

5.1.3   Production

The QGP can be created by heating matter up to a temperature of 2×10^{12} K, which amounts to 175 MeV per particle. This can be accomplished by colliding two large nuclei at high energy (note that 175 MeV is not the energy of the colliding beam). Lead and gold nuclei have been used for such collisions at CERN SPS and BNL RHIC, respectively. The nuclei are accelerated to ultrarelativistic speeds (contracting their length) and directed towards each other, creating a "fireball", in the rare event of a collision. Hydrodynamic simulation predicts this fireball will expand under its own pressure, and cool while expanding. By carefully studying the spherical and elliptic flow, experimentalists put the theory to test.

5.1.4   How the QGP fits into the general scheme of physics

QCD is one part of the modern theory of particle physics called the Standard Model. Other parts of this theory deal with electroweak interactions and neutrinos. The theory of electrodynamics has been tested and found correct to a few parts in a billion. The theory of weak interactions has been tested and found correct to a few parts in a thousand. Perturbative forms of QCD have been tested to a few percent. In contrast, non-perturbative forms of QCD have barely been tested. The study of the QGP is part of this effort to consolidate the grand theory of particle physics.

The study of the QGP is also a testing ground for finite temperature field theory, a branch of theoretical physics which seeks to understand particle physics under conditions of high temperature. Such studies are important to understand the early evolution of our universe: the first hundred microseconds or so. It is crucial to the physics goals of a new generation of observations of the universe (WMAP and its successors). It is also of relevance to Grand Unification Theories which seek to unify the three fundamental forces of nature (excluding gravity).

5.2   Expected properties

5.2.1 Thermodynamics

The cross-over temperature from the normal hadronic to the QGP phase is about 175 MeV. This "crossover" may actually *not* be only a qualitative feature, but instead one may have to do with a true (second order) phase transition, e.g. of the universality class of the three-dimensional Ising model, as some theorists say, e.g. Frithjof Karsch and coworkers from the university of Bielefeld. The phenomena involved correspond to an energy density of a little less than 1 GeV/fm^3. For relativistic matter, pressure and temperature are not independent variables, so the equation of state is a relation between the energy density and the pressure. This has been found through lattice computations, and compared to both perturbation theory and string theory. This is still a matter of active research. Response functions such as the specific heat and various quark number susceptibilities are currently being computed.

5.2.2 Flow

The equation of state is an important input into the flow equations. The speed of sound is currently under investigation in lattice computations. The mean free path of quarks and gluons has been computed using perturbation theory as well as string theory. Lattice computations have been slower here, although the first computations of transport coefficients have recently been concluded. These indicate that the mean free time of quarks and gluons in the QGP may be comparable to the average interparticle spacing: hence the QGP is a liquid as far as its flow properties go. This is very much an active field of research, and these conclusions may evolve rapidly. The incorporation of dissipative phenomena into hydrodynamics is another recent development that is still in an active stage.

5.2.3 Excitation spectrum

Does the QGP really contain (almost) free quarks and gluons? The study of thermodynamic and flow properties would indicate that this is an over-simplification. Many ideas are currently being evolved and will be put to test in the near future. It has been hypothesized recently that some mesons built from heavy quarks do not dissolve until the temperature reaches about 350 MeV. This has led to speculation that many other kinds of bound states may exist in the plasma. Some static properties of the plasma (similar to the Debye screening length) constrain the excitation spectrum.

5.2.4 Glasma hypothesis

Since 2008, there is a discussion about a hypothetical precursor state of the Quark–gluon plasma, the so-called "Glasma", where the dressed particles are condensed into some kind of glassy (or amorphous) state, below the genuine transition between the confined state and the plasma liquid. This would be analogous to the formation of metallic glasses, or amorphous alloys of them, below the genuine onset of the liquid metallic state.

5.3 Experimental situation

Those forms of the QGP that are easiest to compute are not those that are easiest to verify experimentally. While the balance of evidence points towards the QGP being the origin of the detailed properties of the fireball produced in the RHIC, this is the main barrier which prevents experimentalists from declaring a sighting of the QGP. For a summary see 2005 RHIC Assessment.

The important classes of experimental observations are

- Single particle spectra (photons and dileptons)
- Strangeness production
- Photon and muon rates (and J/ψ melting)
- Elliptic flow

- Jet quenching

- Fluctuations

- Hanbury Brown and Twiss effect and Bose–Einstein correlations

In short, a quark–gluon plasma flows like a splat of liquid, and because it's not "transparent" with respect to quarks, it can attenuate jets emitted by collisions. Furthermore, once formed, a ball of quark–gluon plasma, like any hot object, transfers heat internally by radiation. However, unlike in everyday objects, there is enough energy available that gluons (particles mediating the strong force) collide and produce an excess of the heavy (i.e. high-energy) strange quarks. Whereas, if the QGP didn't exist and there was a pure collision, the same energy would be converted into even heavier quarks such as charm quarks or bottom quarks.

5.4   Formation of quark matter

In April 2005, formation of quark matter was tentatively confirmed by results obtained at Brookhaven National Laboratory's Relativistic Heavy Ion Collider (RHIC). The consensus of the four RHIC research groups was that they had created a quark–gluon liquid of very low viscosity. However, contrary to what was at that time still the widespread assumption, it is yet unknown from theoretical predictions whether the QCD "plasma", especially close to the transition temperature, should behave like a gas or liquid. Authors favoring the weakly interacting interpretation derive their assumptions from the lattice QCD calculation, where the entropy density of quark–gluon plasma approaches the weakly interacting limit. However, since both energy density and correlation shows significant deviation from the weakly interacting limit, it has been pointed out by many authors that there is in fact no reason to assume a QCD "plasma" close to the transition point should be weakly interacting, like electromagnetic plasma (see, e.g.,[15]). That being said, systematically improvable perturbative QCD quasiparticle models do a very good job of reproducing the lattice data for thermodynamical observables (pressure, entropy, quark susceptibility), including the aforementioned "significant deviation from the weakly interacting limit", down to temperatures on the order of 2 to 3 times the critical temperature for the transition.[16][17][18]

5.5   See also

- Hadrons (that is mesons and baryons) and confinement

- Hadronization

- List of plasma (physics) articles

- Neutron stars

- Plasma physics

- QCD matter Quantum Chromodynamics matter

- Quantum electrodynamics

- Quantum chromodynamics

- Quantum hydrodynamics

- Relativistic plasma

- Relativistic nuclear collision

- Strangeness production

- Strange matter

- Color-glass condensate

5.6 References

[1] Bohr, Henrik; Nielsen, H. B. (1977). "Hadron production from a boiling quark soup: quark model predicting particle ratios in hadronic collisions". *Nuclear Physics B* **128** (2): 275. Bibcode:1977NuPhB.128..275B. doi:10.1016/0550-3213(77)90032-3.

[2] LHC creates liquid from Big Bang

[3] Quark-gluon plasma goes liquid - physicsworld.com

[4] A New State of Matter - Experiments

[5] Relativistic Heavy Ion Collider, RHIC

[6] http://www.bnl.gov/rhic/news2/news.asp?a=1074&t=pr 'Perfect' Liquid Hot Enough to be Quark Soup

[7] Alice Experiment: Welcome to ALICE Portal

[8] CERN Press Release November 4th 2010

[9] Hot stuff: CERN physicists create record-breaking subatomic soup : Nature News Blog

[10] The Indian Lattice Gauge Theory Initiative

[11] WA Zajc (2008). "The fluid nature of quark-gluon plasma".*Nuclear Physics A***805**: 283c–294c.arXiv:0802.3552.Bibcode:2 doi:10.1016/j.nuclphysa.2008.02.285.

[12] Plümer, M.; Raha, S. & Weiner, R. M. (1984). "How free is the quark-gluon plasma". *Nucl. Phys. A* **418**: 549–557. Bibcode:1984NuPhA.418..549P. doi:10.1016/0375-9474(84)90575-X..

[13] Plümer, M.; Raha, S. & Weiner, R. M. (1984). "Effect of confinement on the sound velocity in a quark-gluon plasma". *Phys. Lett. B* **139** (3): 198–202. Bibcode:1984PhLB..139..198P. doi:10.1016/0370-2693(84)91244-9..

[14] Lattice-QCD calculations of the Quark-Gluon Plasma have been reviewed in and in

[15] Miklos Gyulassy (2004). "The QGP Discovered at RHIC". arXiv:nucl-th/0403032 [nucl-th].

[16] Andersen; Leganger; Strickland; Su (2011). "NNLO hard-thermal-loop thermodynamics for QCD". *Physics Letters B* **696** (5): 468. arXiv:1009.4644. Bibcode:2011PhLB..696..468A. doi:10.1016/j.physletb.2010.12.070.

[17] Andersen; Michael Strickland; Nan Su (2010). "Gluon Thermodynamics at Intermediate Coupling". *Physical Review Letters* **104** (12). arXiv:0911.0676. Bibcode:2010PhRvL.104l2003A. doi:10.1103/PhysRevLett.104.122003.

[18] Blaizot; Iancu; Rebhan (2003). "Thermodynamics of the high-temperature quark-gluon plasma". arXiv:hep-ph/0303185 [hep-ph].

5.7 External links

- The Relativistic Heavy Ion Collider at Brookhaven National Laboratory

- The Alice Experiment at CERN

- The Indian Lattice Gauge Theory Initiative

- Quark matter reviews: 2004 theory, 2004 experiment

- Quark-Gluon Plasma reviews: 2011 theory

- Lattice reviews: 2003, 2005

- BBC article mentioning Brookhaven results (2005)

- Physics News Update article on the quark-gluon liquid, with links to preprints

- Read for free : "Hadrons and Quark-Gluon Plasma" by Jean Letessier and Johann Rafelski Cambridge University Press (2002) ISBN 0-521-38536-9, Cambridge, UK;

Chapter 6

Glueball

In particle physics, a **glueball** is a hypothetical composite particle.[1] It consists solely of gluon particles, without valence quarks. Such a state is possible because gluons carry color charge and experience the strong interaction. Glueballs are extremely difficult to identify in particle accelerators, because they mix with ordinary meson states.[2]

Theoretical calculations show that glueballs should exist at energy ranges accessible with current collider technology. However, due to the aforementioned difficulty (among others), they have (as of 2013) so far not been observed and identified with certainty.[3] The prediction that glueballs exist is one of the most important predictions of the Standard Model of particle physics that has not yet been confirmed experimentally.[4]

6.1 Properties of glueballs

In principle, it is theoretically possible for all properties of glueballs to be calculated exactly and derived directly from the equations and fundamental physical constants of quantum chromodynamics (QCD) without further experimental input. So, the predicted properties of these hypothetical particles can be described in exquisite detail using only Standard Model physics which have wide acceptance in the theoretical physics literature. But, the fact that QCD calculations are so difficult that solutions to these equations are almost always numerical approximations (reached by several very different methodologies) and the considerable uncertainty in the measurement of some of the relevant key physical constants can lead to variation in theoretical predictions of glueball properties like mass and branching ratios in glueball decays.

6.1.1 Constituent particles and color charge

Theoretical studies of glueballs have focused on glueballs consisting of either two gluons or three gluons, by analogy to mesons and baryons that have two and three quarks respectively. As in the case of mesons and baryons, glueballs would be QCD color charge neutral (aka isospin = 0). The baryon number of a glueball is zero.

6.1.2 Total angular momentum

Two gluon glueballs can have total angular momentum (J) of 0 (which are scalar or pseudo-scalar) or 2 (tensor). Three gluon glueballs can have total angular momentum (J) of 1 (vector boson) or 3. All glueballs have integer total angular momentum which implies that they are bosons rather than fermions.

Glueballs are the only particles predicted by the Standard Model with total angular momentum (J) (sometimes called "intrinsic spin") that could be either 2 or 3 in their ground states, although mesons made of two quarks with J=0 and J=1 with similar masses have been observed and excited states of other mesons can have these values of total angular momentum.

Fundamental particles with ground states having J=0 or J=2 are easily distinguished from glueballs. The hypothetical graviton, while having a total angular momentum J=2 would be massless and lack color charge, and so would be easily distinguished from glueballs. The Standard Model Higgs boson for which an experimentally measured mass of about 125-126 GeV/c^2 has been determined (although the status of the measured particle as a true Standard Model Higgs boson has not been definitively established), is the only fundamental particle with J=0 in the Standard Model, also lacks color charge and hence does not engage in strong force interactions. The Higgs boson is about 25-80 times as heavy as the mass of the various glueball states predicted by the Standard Model.

6.1.3 Electric charge

All glueballs would have electric charge, Q(e), of zero as gluons themselves do not have an electric charge.

6.1.4 Mass and parity

Glueballs are predicted by quantum chromodynamics to be massive, notwithstanding the fact that gluons themselves have zero rest mass in the Standard Model. Glueballs with all four possible combinations of quantum numbers P (parity) and C (c-parity) for every possible total angular momentum have been considered, producing at least fifteen possible glueball states including excited glueball states that share the same quantum numbers but have differing masses with the lightest states having masses as low as 1.4 GeV/c^2 (for a glueball with quantum numbers J=0, P=+, C=+), and the heaviest states having masses as great as almost 5 GeV/c^2 (for a glueball with quantum numbers J=0, P=+, C=-).[5]

These masses are on the same order of magnitude as the masses of many experimentally observed mesons and baryons, as well as to the masses of the tau lepton, charm quark, bottom quark, some hydrogen isotopes, and some helium isotopes.

6.1.5 Stability and decay channels

Just as all Standard Model mesons and baryons, except the proton, are unstable in isolation, all glueballs are predicted by the Standard Model to be unstable in isolation, with various QCD calculations predicting the total decay width (which is functionally related to half-life) for various glueball states. QCD calculations also make predictions regarding the expected decay patterns of glueballs.[6][7] For example, glueballs would not have radiative or two photon decays, but would have decays into pairs of pions, pairs of kaons, or pairs of eta mesons.[6]

6.2 Practical impact on macroscopic low energy physics

Because Standard Model glueballs are so ephemeral (decaying almost immediately into more stable decay products) and are only generated in high energy physics, glueballs only arise synthetically in the natural conditions found on Earth that humans can easily observe. They are scientifically notable mostly because they are a testable prediction of the Standard Model, and not because of phenomenological impact on macroscopic processes, or their engineering applications.

6.3 Lattice QCD simulations

Lattice field theory provides a way to study the glueball spectrum theoretically and from first principles. Some of the first quantities calculated using lattice QCD methods (in 1980) were glueball mass estimates.[9] Morningstar and Peardon[10] computed in 1999 the masses of the lightest glueballs in QCD without dynamical quarks. The three lowest states are tabulated below. The presence of dynamical quarks would slightly alter these data, but also makes the computations more difficult. Since that time calculations within QCD (lattice and sum rules) find the lightest glueball to be a scalar with mass in the range of about 1000–1700 MeV.[11]

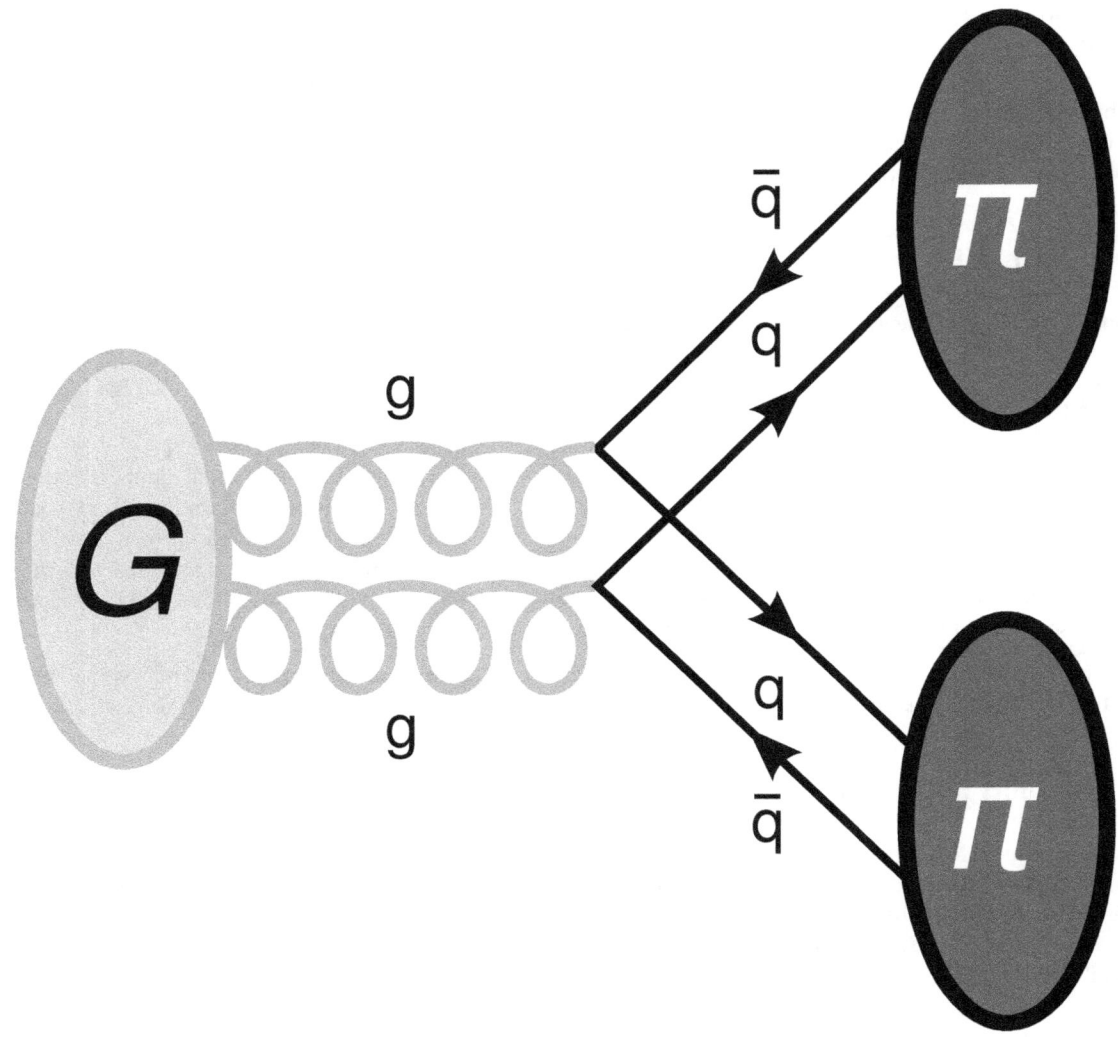

Feynman diagram of a glueball (G) decaying to two pions (π). Such decays help the study of and search for glueballs.[8]

6.4   Experimental candidates

Particle accelerator experiments are often able to identify unstable composite particles and assign masses to those particles to a precision of approximately 10 MeV/c^2, without being able to immediately assign to the particle resonance that is observed all of the properties of that particle. Scores of such particles have been detected, although particles detected in some experiments but not others can be viewed as doubtful. Some of the candidate particle resonances that could be glueballs, although the evidence is not definitive, include the following:

Vector, Pseudo-Vector, or Tensor Glueball Candidates:

- X(3020) observed by the BaBar collaboration is a candidate for an excited state of the 2-+, 1+- or 1-- glueball states with a mass of about 3.02 GeV/c^2.[4]

Scalar Glueball Candidates:

- $f_0(500)$ also known as σ -- the properties of this particle are possibly consistent with a 1000 MeV or 1500 MeV mass glueball.[12]

- $f_0(980)$ -- the structure of this composite particle is consistent with the existence of a light glueball.[12]

- $f_0(1370)$ -- existence of this resonance is disputed but is a candidate for a glueball-meson mixing state[12]

- $f_0(1500)$ -- existence of this resonance is undisputed but its status as a glueball-meson mixing state or pure glueball is not well established.[12]

- $f_0(1710)$ -- existence of this resonance is undisputed but its status as a glueball-meson mixing state or pure glueball is not well established.[12]

Other Glueball Candidates:

- Gluon jets at the LEP experiment show a 40% excess over theoretical expectations of electromagnetically neutral clusters which suggests that electromagnetically neutral particles expected in gluon rich environments such as glueballs are likely to be present.[12]

Many of these candidates have been the subject of active investigation for at least eighteen years.[6] The GlueX experiment, scheduled to begin in 2014, has been specifically designed to produce more definitive experimental evidence glueballs.[13]

6.5   See also

- Exotic meson

- GlueX

- Gluon

- Yang–Mills theory

6.6   References

[1]   • Frank Close and Phillip R. Page, "Glueballs", *Scientific American*, vol. 279 no. 5 (November 1998) pp. 80–85

[2] Vincent Mathieu; Nikolai Kochelev; Vicente Vento (2009). "The Physics of Glueballs". *International Journal of Modern Physics E* **18**: 1–49. arXiv:0810.4453. Bibcode:2009IJMPE..18....1M. doi:10.1142/S0218301309012124. Glueball on arxiv.org

[3] Wolfgang Ochs, "The Status of Glueballs" J.Phys.G: Nuclear and Particle Physics 40, 67 (2013) DOI: 10.1088/0954-3899/40/4/0 http://arxiv.org/pdf/1301.5183v3.pdf

[4] Y.K. Hsiao, C.Q. Geng, "Identifying Glueball at 3.02 GeV in Baryonic B Decays" (Version 2: October 9, 2013) http://arxiv.org/abs/1302.3331

[5] Wolfgang Ochs, "The Status of Glueballs" J.Phys.G: Nuclear and Particle Physics 40, 6 (2013) DOI: 10.1088/0954-3899/40/4/04 http://arxiv.org/pdf/1301.5183v3.pdf

[6] Walter Taki, "Search for Glueballs" (1996) http://www.slac.stanford.edu/cgi-wrap/getdoc/ssi96-006.pdf

[7] See, e.g., Walaa I. Eshraim, Stanislaus Janowski, "Branching ratios of the pseudoscalar glueball with a mass of 2.6 GeV", prepared for Proceedings of Confinement X - Conference on Quark Confinement and the Hadron Spectrum (Munich/Germany, 8–12 October 2012) (pre-print published January 15, 2013) http://arxiv.org/abs/1301.3345

[8] T. Cohen, F. J. Llanes-Estrada, J. R. Pelaez, J. Ruiz de Elvira (2014). "Non-ordinary light meson couplings and the 1/Nc expansion". arXiv:1405.4831 [hep-ph].

[9] B. Berg. Plaquette-plaquette correlations in the su(2) lattice gauge theory. Phys. Lett., B97:401, 1980.

[10] Colin J. Morningstar; Mike Peardon (1999). "Glueball spectrum from an anisotropic lattice study". *Physical Review D* **60** (3): 034509. arXiv:hep-lat/9901004. Bibcode:1999PhRvD..60c4509M. doi:10.1103/PhysRevD.60.034509.

[11] Wolfgang Ochs, "The status of glueballs" Source: JOURNAL OF PHYSICS G-NUCLEAR AND PARTICLE PHYSICS Volume: 40 Issue: 4 Article Number: 043001 DOI: 10.1088/0954-3899/40/4/043001 Published: APR 2013

[12] Wolfgang Ochs(2013). "The status of glueballs".*Journal of Physics G***40**(4): 043001.arXiv:1301.5183.Bibcode:2013JPhG... doi:10.1088/0954-3899/40/4/043001.

[13] "The Physics of GlueX".

Chapter 7

List of mesons

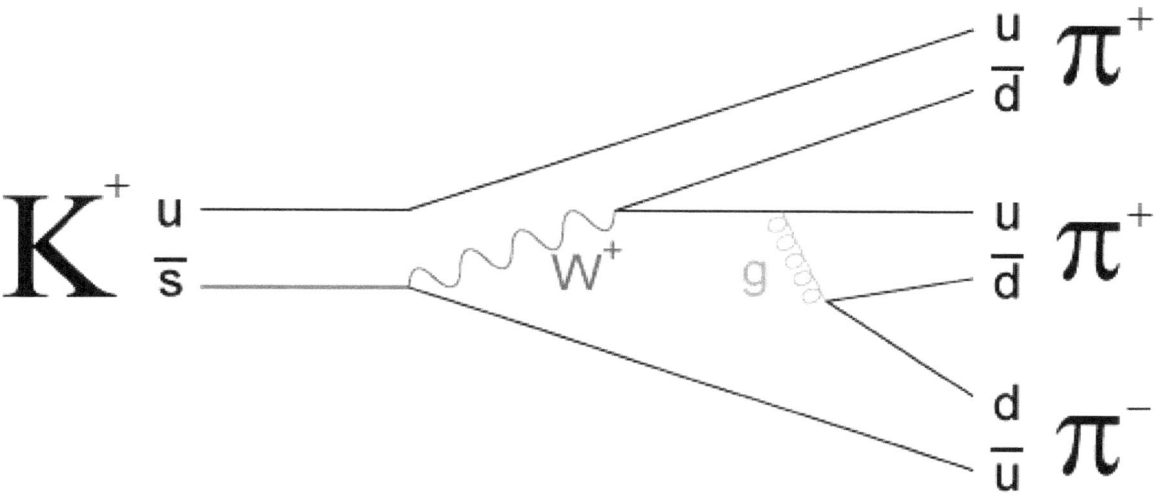

The decay of a kaon (K+) into three pions (2 π+, 1 π−) is a process that involves both weak and strong interactions.
Weak interactions: The strange antiquark (s) of the kaon transmutes into an up antiquark (u) by the emission of a W+ boson; the W+
boson subsequently decays into a down antiquark (d) and an up quark (u).
Strong interactions: An up quark (u) emits a gluon (g) which decays into a down quark (d) and a down antiquark (d).

This list is of all known and predicted scalar, pseudoscalar and vector mesons. See list of particles for a more detailed list of particles found in particle physics.

Mesons are unstable subatomic particles composed of one quark and one antiquark. They are part of the hadron particle family – particles made of quarks. The other members of the hadron family are the baryons – subatomic particles composed of three quarks. The main difference between mesons and baryons is that mesons have integer spin (thus are bosons) while baryons are fermions (half-integer spin). Because mesons are bosons, the Pauli exclusion principle does not apply to them. Because of this, they can act as force mediating particles on short distances, and thus play a part in processes such as the nuclear interaction.

Since mesons are composed of quarks, they participate in both the weak and strong interactions. Mesons with net electric charge also participate in the electromagnetic interaction. They are classified according to their quark content, total angular momentum, parity, and various other properties such as C-parity and G-parity. While no meson is stable, those of lower mass are nonetheless more stable than the most massive mesons, and are easier to observe and study in particle accelerators or in cosmic ray experiments. They are also typically less massive than baryons, meaning that they are more easily produced in experiments, and will exhibit higher-energy phenomena sooner than baryons would. For example, the charm quark was first seen in the J/Psi meson (J/ψ) in 1974,[1][2] and the bottom quark in the upsilon meson (Υ) in 1977.[3]

Each meson has a corresponding antiparticle (antimeson) where quarks are replaced by their corresponding antiquarks and vice versa. For example, a positive pion (π+) is made of one up quark and one down antiquark; and its corresponding antiparticle, the negative pion (π−), is made of one up antiquark and one down quark. Some experiments show the evidence of *tetraquarks* – "exotic" mesons made of two quarks and two antiquarks, but the particle physics community as a whole does not view their existence as likely, although still possible.[4]

The symbols encountered in these lists are: I (*isospin*), J (*total angular momentum*), P (*parity*), C (*C-parity*), G (*G-parity*), u (*up quark*), d (*down quark*), s (*strange quark*), c (*charm quark*), b (*bottom quark*), Q (*charge*), B (*baryon number*), S (*strangeness*), C (*charm*), and B′ (*bottomness*), as well as a wide array of subatomic particles (hover for name).

7.1   Summary table

Because this table was initially derived from published results and many of those results were preliminary, as many as 64 of the mesons in the following table may not exist or have the wrong mass or quantum numbers.

7.2   Meson properties

The following lists detail all known and predicted pseudoscalar ($J^P = 0^-$) and vector ($J^P = 1^-$) mesons.

The properties and quark content of the particles are tabulated below; for the corresponding antiparticles, simply change quarks into antiquarks (and vice versa) and flip the sign of Q, B, S, C, and B′. Particles with † next to their names have been predicted by the standard model but not yet observed. Values in red have not been firmly established by experiments, but are predicted by the quark model and are consistent with the measurements.

7.2.1   Pseudoscalar mesons

[a] ∧ Makeup inexact due to non-zero quark masses.

[b] ∧ PDG reports the resonance width (Γ). Here the conversion $\tau = {}^\hbar\!/\Gamma$ is given instead.

[c] ∧ Strong eigenstate. No definite lifetime (see kaon notes below)

[d] ∧ The mass of the K0
L and K0
S are given as that of the K0. However, it is known that a difference between the masses of the K0
L and K0
S on the order of 2.2×10^{-11} MeV/c^2 exists.[15]

[e] ∧ Weak eigenstate. Makeup is missing small CP–violating term (see notes on neutral kaons below).

7.2.2   Vector mesons

[f] ∧ PDG reports the resonance width (Γ). Here the conversion $\tau = {}^\hbar\!/\Gamma$ is given instead.

[g] ∧ The exact value depends on the method used. See the given reference for detail.

7.2.3   Notes on neutral kaons

There are two complications with neutral kaons:[34]

- Due to neutral kaon mixing, the K0
S and K0

L are not eigenstates of strangeness. However, they *are* eigenstates of the weak force, which determines how they decay, so these are the particles with definite lifetime.

- The linear combinations given in the table for the K0
  S and K0
  L are not exactly correct, since there is a small correction due to CP violation. See CP violation in kaons.

Note that these issues also exist in principle for other neutral flavored mesons; however, the weak eigenstates are considered separate particles only for kaons because of their dramatically different lifetimes.[34]

7.3 See also

- List of baryons
- List of particles
- Timeline of particle discoveries

7.4 References

[1] J.J. Aubert *et al.* (1974)

[2] J.E. Augustin *et al.* (1974)

[3] S.W. Herb *et al.* (1977)

[4] C. Amsler *et al.* (2008): Charmonium States

[5] K.A. Olive *et al.* (2014): Meson Summary Table

[6] K.A. Olive *et al.* (2014): Particle listings – π±

[7] K.A. Olive *et al.* (2014): Particle listings – π0

[8] K.A. Olive *et al.* (2014): Particle listings – η

[9] K.A. Olive *et al.* (2014): Particle listings – η′

[10] K.A. Olive *et al.* (2014): Particle listings – η
     c

[11] K.A. Olive *et al.* (2014): Particle listings – η
     b

[12] K.A. Olive *et al.* (2014): Particle listings – K±

[13] K.A. Olive *et al.* (2014): Particle listings – K0

[14] K.A. Olive *et al.* (2014): Particle listings – K0
     S

[15] K.A. Olive *et al.* (2014): Particle listings – K0
     L

[16] K.A. Olive *et al.* (2014): Particle listings – D±

[17] K.A. Olive *et al.* (2014): Particle listings – D0

[18] K.A. Olive *et al.* (2014): Particle listings – D±
     s

[19] K.A. Olive *et al.* (2014): Particle listings – B±

[20] K.A. Olive *et al.* (2014): Particle listings – B0

[21] K.A. Olive *et al.* (2014): Particle listings – B0
     s

[22] K.A. Olive *et al.* (2014): Particle listings – B±
     c

[23] K.A. Olive *et al.* (2014): Particle listings – ρ

[24] K.A. Olive *et al.* (2014): Particle listings – ω(782)

[25] K.A. Olive *et al.* (2014): Particle listings – φ

[26] K.A. Olive *et al.* (2014): Particle listings – J/Ψ

[27] K.A. Olive *et al.* (2014): Particle listings – ϒ(1S)

[28] K.A. Olive *et al.* (2014): Particle listings – K∗(892)

[29] K.A. Olive *et al.* (2014): Particle listings – D∗±(2010)

[30] K.A. Olive *et al.* (2014): Particle listings – D∗0(2007)

[31] K.A. Olive *et al.* (2014): Particle listings – D∗±
     s

[32] K.A. Olive *et al.* (2014): Particle listings – B∗

[33] K.A. Olive *et al.* (2014): Particle listings – B∗
     s

[34] J.W. Cronin (1980)

7.4.1 Bibliography

- K.A. Olive *et al.* (Particle Data Group) (2014). "Review of Particle Physics". *Chinese Physics C* **38** (9): 090001.

- M.S. Sozzi (2008a). "Parity". *Discrete Symmetries and CP Violation: From Experiment to Theory*. Oxford University Press. pp. 15–87. ISBN 0-19-929666-9.

- M.S. Sozzi (2008a). "Charge Conjugation". *Discrete Symmetries and CP Violation: From Experiment to Theory*. Oxford University Press. pp. 88–120. ISBN 0-19-929666-9.

- M.S. Sozzi (2008c). "CP-Symmetry". *Discrete Symmetries and CP Violation: From Experiment to Theory*. Oxford University Press. pp. 231–275. ISBN 0-19-929666-9.

- C. Amsler *et al.* (Particle Data Group); Amsler; Doser; Antonelli; Asner; Babu; Baer; Band; Barnett; Bergren; Beringer; Bernardi; Bertl; Bichsel; Biebel; Bloch; Blucher; Blusk; Cahn; Carena; Caso; Ceccucci; Chakraborty; Chen; Chivukula; Cowan; Dahl; d'Ambrosio; Damour et al. (2008). "Review of Particle Physics". *Physics Letters B* **667** (1): 1–1340. Bibcode:2008PhLB..667....1P. doi:10.1016/j.physletb.2008.07.018.

- S.S.M. Wong (1998). "Nucleon Structure". *Introductory Nuclear Physics* (2nd ed.). John Wiley & Sons. pp. 21–56. ISBN 0-471-23973-9.

- R. Shankar (1994). *Principles of Quantum Mechanics* (2nd ed.). Plenum Press. ISBN 0-306-44790-8.

- K. Gottfried, V.F. Weisskopf (1986). "Hadronic Spectroscopy: G-parity". *Concepts of Particle Physics* **2**. Oxford University Press. pp. 303–311. ISBN 0-19-503393-0.

- J.W. Cronin (1980). "CP Symmetry Violation – The Search for its origin" (PDF). *Nobel Lecture*. The Nobel Foundation.

- V.L. Fitch (1980). "The Discovery of Charge – Conjugation Parity Asymmetry" (PDF). *Nobel Lecture*. The Nobel Foundation.

- S.W. Herb; Hom, D.; Lederman, L.; Sens, J.; Snyder, H.; Yoh, J.; Appel, J.; Brown, B. et al. (1977). "Observation of a Dimuon Resonance at 9.5 Gev in 400-GeV Proton-Nucleus Collisions". *Physical Review Letters* **39** (5): 252–255. Bibcode:1977PhRvL..39..252H. doi:10.1103/PhysRevLett.39.252.

- J.J. Aubert; Becker, U.; Biggs, P.; Burger, J.; Chen, M.; Everhart, G.; Goldhagen, P.; Leong, J. et al. (1974). "Experimental Observation of a Heavy Particle J". *Physical Review Letters* **33** (23): 1404–1406. Bibcode:1974PhRvL.. 33.1404A.doi:10.1103/PhysRevLett.33.1404.

- J.E. Augustin; Boyarski, A.; Breidenbach, M.; Bulos, F.; Dakin, J.; Feldman, G.; Fischer, G.; Fryberger, D. et al. (1974). "Discovery of a Narrow Resonance in e⁺e⁻ Annihilation". *Physical Review Letters* **33** (23): 1406–1408. Bibcode:1974PhRvL..33.1406A. doi:10.1103/PhysRevLett.33.1406.

- M. Gell-Mann (1964). "A Schematic of Baryons and Mesons". *Physics Letters* **8**(3): 214–215.Bibcode:1964PhL... doi:10.1016/S0031-9163(64)92001-3.

- E. Wigner (1937). "On the Consequences of the Symmetry of the Nuclear Hamiltonian on the Spectroscopy of Nuclei". *Physical Review* **51** (2): 106–119. Bibcode:1937PhRv...51..106W. doi:10.1103/PhysRev.51.106.

- W. Heisenberg (1932). "Über den Bau der Atomkerne I". *Zeitschrift für Physik* (in German) **77** (1–2): 1–11. Bibcode:1932ZPhy...77....1H. doi:10.1007/BF01342433.

- W. Heisenberg (1932). "Über den Bau der Atomkerne II". *Zeitschrift für Physik* (in German) **78** (3–4): 156–164. Bibcode:1932ZPhy...78..156H. doi:10.1007/BF01337585.

- W. Heisenberg (1932). "Über den Bau der Atomkerne III". *Zeitschrift für Physik* (in German) **80** (9–10): 587–596. Bibcode:1933ZPhy...80..587H. doi:10.1007/BF01335696.

7.5 External links

- Particle Data Group – The Review of Particle Physics (2008)

- Mesons made thinkable, an interactive visualisation allowing physical properties to be compared

Chapter 8

Timeline of particle discoveries

This is a **timeline of subatomic particle discoveries**, including all particles thus far discovered which appear to be elementary (that is, indivisible) given the best available evidence. It also includes the discovery of composite particles and antiparticles that were of particular historical importance.

More specifically, the inclusion criteria are:

- Elementary particles from the Standard Model of particle physics that have so far been observed. The Standard Model is the most comprehensive existing model of particle behavior. All Standard Model particles including the Higgs boson have been verified, and all other observed particles are combinations of two or more Standard Model particles.

- Antiparticles which were historically important to the development of particle physics, specifically the positron and antiproton. The discovery of these particles required very different experimental methods from that of their ordinary matter counterparts, and provided evidence that *all* particles had antiparticles—an idea that is fundamental to quantum field theory, the modern mathematical framework for particle physics. In the case of most subsequent particle discoveries, the particle and its anti-particle were discovered essentially simultaneously.

- Composite particles which were the first particle discovered containing a particular elementary constituent, or whose discovery was critical to the understanding of particle physics.

8.1 See also

- List of mesons

- List of baryons

- List of particles

- physics

8.2 References

[1] Hockberger, P. E. (2002). "A history of ultraviolet photobiology for humans, animals and microorganisms". *Photochem. Photobiol.* **76** (6): 561–579. doi:10.1562/0031-8655(2002)0760561AHOUPF2.0.CO2. ISSN 0031-8655. PMID 12511035.

[2]The ozone layer protects humans from this. Lyman, T. (1914). "Victor Schumann".*Astrophysical Journal***38**: 1–4.Bibcode: doi:10.1086/142050.

[3] W.C. Röntgen (1895). "Über ein neue Art von Strahlen. Vorlaufige Mitteilung". *Sitzber. Physik. Med. Ges.* **137**: 1. as translated in A. Stanton (1896). "On a New Kind of Rays". *Nature* **53** (1369): 274–276. Bibcode:1896Natur..53R.274.. doi:10.1038/053274b0.

[4] J.J. Thomson (1897). "Cathode Rays". *Philosophical Magazine* **44** (269): 293–316. doi:10.1080/14786449708621070.

[5] E. Rutherford (1899). "Uranium Radiation and the Electrical Conduction Produced by it". *Philosophical Magazine* **47** (284): 109–163. doi:10.1080/14786449908621245.

[6] P. Villard (1900). "Sur la Réflexion et la Réfraction des Rayons Cathodiques et des Rayons Déviables du Radium". *Comptes Rendus de l'Académie des Sciences* **130**: 1010.

[7] E. Rutherford (1911). "The Scattering of α- and β- Particles by Matter and the Structure of the Atom". *Philosophical Magazine* **21** (125): 669–688. doi:10.1080/14786440508637080.

[8] E. Rutherford (1919). "Collision of α Particles with Light Atoms IV. An Anomalous Effect in Nitrogen". *Philosophical Magazine* **37**: 581.

[9] J. Chadwick (1932). "Possible Existence of a Neutron".*Nature***129**(3252): 312.Bibcode:1932Natur.129Q.312C.doi:10.1038/

[10] E. Rutherford (1920). "Nuclear Constitution of Atoms".*Proceedings of the Royal Society A***97**(686): 374–400.Bibcode:1920 doi:10.1098/rspa.1920.0040.

[11] C.D. Anderson (1932). "The Apparent Existence of Easily Deflectable Positives".*Science***76**(1967): 238–9.Bibcode:1932Sc doi:10.1126/science.76.1967.238. PMID 17731542.

[12] S.H. Neddermeyer, C.D. Anderson (1937). "Note on the nature of Cosmic-Ray Particles". *Physical Review* **51** (10): 884–886. Bibcode:1937PhRv...51..884N. doi:10.1103/PhysRev.51.884.

[13] M. Conversi, E. Pancini, O. Piccioni (1947). "On the Disintegration of Negative Muons". *Physical Review* **71** (3): 209–210. Bibcode:1947PhRv...71..209C. doi:10.1103/PhysRev.71.209.

[14] C.D. Anderson (1935). "On the Interaction of Elementary Particles". *Proceedings of the Physico-Mathematical Society of Japan* **17**: 48.

[15] G.D. Rochester, C.C. Butler (1947). "Evidence for the Existence of New Unstable Elementary Particles". *Nature* **160** (4077): 855–857. Bibcode:1947Natur.160..855R. doi:10.1038/160855a0.

[16] The Strange Quark

[17] O. Chamberlain, E. Segrè, C. Wiegand, T. Ypsilantis (1955). "Observation of Antiprotons". *Physical Review* **100** (3): 947–950. Bibcode:1955PhRv..100..947C. doi:10.1103/PhysRev.100.947.

[18] F. Reines, C.L. Cowan (1956). "The Neutrino".*Nature***178**(4531): 446–449.Bibcode:1956Natur.178..446R.doi:10.1038/1

[19] G. Danby; et al. (1962). "Observation of High-Energy Neutrino Reactions and the Existence of Two Kinds of Neutrinos". *Physical Review Letters* **9** (1): 36–44. Bibcode:1962PhRvL...9...36D. doi:10.1103/PhysRevLett.9.36.

[20] R. Nave. "The Xi Baryon". Hyperphysics. Retrieved 20 June 2009.

[21] E.D. Bloom; et al. (1969). "High-Energy Inelastic *e–p* Scattering at 6° and 10°". *Physical Review Letters* **23** (16): 930–934. Bibcode:1969PhRvL..23..930B. doi:10.1103/PhysRevLett.23.930.

[22] M. Breidenbach; et al. (1969). "Observed Behavior of Highly Inelastic Electron-Proton Scattering". *Physical Review Letters* **23** (16): 935–939. Bibcode:1969PhRvL..23..935B. doi:10.1103/PhysRevLett.23.935.

[23] J.J. Aubert; et al. (1974). "Experimental Observation of a Heavy Particle *J*". *Physical Review Letters* **33** (23): 1404–1406. Bibcode:1974PhRvL.33.1404A. doi:10.1103/PhysRevLett.33.1404.

[24] J.-E. Augustin; et al. (1974). "Discovery of a Narrow Resonance in e^+e^- Annihilation". *Physical Review Letters* **33** (23): 1406–1408. Bibcode:1974PhRvL.33.1406A. doi:10.1103/PhysRevLett.33.1406.

[25] B.J. Bjørken, S.L. Glashow (1964). "Elementary Particles and SU(4)".*Physics Letters***11**(3): 255–257.Bibcode:1964PhL....11 doi:10.1016/0031-9163(64)90433-0.

[26] M.L. Perl; et al. (1975). "Evidence for Anomalous Lepton Production in e^+-e^- Annihilation". *Physical Review Letters* **35** (22): 1489–1492. Bibcode:1975PhRvL..35.1489P. doi:10.1103/PhysRevLett.35.1489.

[27] S.W. Herb; et al. (1977). "Observation of a Dimuon Resonance at 9.5 GeV in 400-GeV Proton-Nucleus Collisions". *Physical Review Letters* **39** (5): 252–255. Bibcode:1977PhRvL..39..252H. doi:10.1103/PhysRevLett.39.252.

[28] D.P. Barber; et al. (1979). "Discovery of Three-Jet Events and a Test of Quantum Chromodynamics at PETRA". *Physical Review Letters* **43** (12): 830–833. Bibcode:1979PhRvL..43..830B. doi:10.1103/PhysRevLett.43.830.

[29] J.J. Aubert *et al.* (European Muon Collaboration) (1983). "The ratio of the nucleon structure functions F_2^N for iron and deuterium". *Physics Letters B* **123** (3–4): 275–278. Bibcode:1983PhLB..123..275A. doi:10.1016/0370-2693(83)90437-9.

[30] G. Arnison *et al.* (UA1 collaboration) (1983). "Experimental observation of lepton pairs of invariant mass around 95 GeV/c^2 at the CERN SPS collider". *Physics Letters B* **126** (5): 398–410. Bibcode:1983PhLB..126..398A. doi:10.1016/0370-2693(83)90188-0.

[31] F. Abe *et al.* (CDF collaboration) (1995). "Observation of Top quark production in p–p Collisions with the Collider Detector at Fermilab". *Physical Review Letters* **74** (14): 2626–2631. arXiv:hep-ex/9503002. Bibcode:1995PhRvL..74.2626A. doi:10.1103/PhysRevLett.74.2626. PMID 10057978.

[32] S. Arabuchi *et al.* (D0 collaboration) (1995). "Observation of the Top Quark". *Physical Review Letters* **74** (14): 2632–2637. arXiv:hep-ex/9503003. Bibcode:1995PhRvL..74.2632A. doi:10.1103/PhysRevLett.74.2632. PMID 10057979.

[33] G. Baur; et al. (1996). "Production of Antihydrogen". *Physics Letters B* **368** (3): 251–258. Bibcode:1996PhLB..368..251B. doi:10.1016/0370-2693(96)00005-6.

[34] "Physicists Find First Direct Evidence for Tau Neutrino at Fermilab" (Press release). Fermilab. 20 July 2000. Retrieved 20 March 2010.

[35] Boyle, Alan (4 July 2012). "Milestone in Higgs quest: Scientists find new particle". *MSNBC* (MSNBC). Retrieved 5 July 2012.

- V.V. Ezhela; et al. (1996). *Particle Physics: One Hundred Years of Discoveries: An Annotated Chronological Bibliography*. Springer–Verlag. ISBN 1-56396-642-5.

8.3 Text and image sources, contributors, and licenses

8.3.1 Text

- **Gluon** *Source:* https://en.wikipedia.org/wiki/Gluon?oldid=681546689 *Contributors:* AxelBoldt, CYD, Bryan Derksen, Gdarin, TakuyaMurata, Card~enwiki, Looxix~enwiki, Ellywa, Ahoerstemeier, Med, Schneelocke, Phys, Phil Boswell, Donarreiskoffer, Fredrik, Merovingian, Hadal, Giftlite, Herbee, Xerxes314, Eequor, Darrien, Keith Edkins, RetiredUser2, Icairns, Mike Rosoft, AlexChurchill, HedgeHog, Kenny TM~~enwiki, David Schaich, Ioliver, Mashford, El C, Kwamikagami, Ardric47, Obradovic Goran, Alansohn, Guy Harris, Dachannien, Ricky81682, Batmanand, Velella, Kazvorpal, April Arcus, Forteblast, Mpatel, Palica, BD2412, Kbdank71, Rjwilmsi, Macumba, Strait, Mike Peel, Bubba73, Klortho, FlaBot, Srleffler, Chobot, YurikBot, Wavelength, Bambaiah, Hairy Dude, Jimp, JabberWok, Zelmerszoetrop, Salsb, SCZenz, Randolf Richardson, Ravedave, Danlaycock, Bota47, LeonardoRob0t, Anclation~enwiki, Physicsdavid, Erudy, GrinBot~enwiki, Kgf0, SmackBot, Melchoir, Cessator, Benjaminevans82, Abtal, MK8, Colonies Chris, Can't sleep, clown will eat me, Decltype, Qcdmaestro, Edconrad, Darkpoison99, FredrickS, Omsharan, Pegasusbot, Gregbard, ProfessorPaul, Thijs!bot, Headbomb, Rriegs, Oreo Priest, AntiVandalBot, Shambolic Entity, Deflective, Mujokan, Yill577, Happycool, Mother.earth, Martynas Patasius, WiiWillieWiki, HEL, Hans Dunkelberg, Gombang, Inwind, Sheliak, Jonthaler, VolkovBot, TXiKiBoT, Davehi1, Kriak, Anonymous Dissident, Imasleepviking, AlleborgoBot, EJF, SieBot, Steven Crossin, OKBot, ClueBot, Wwheaton, Qsaw, Nucularphysicist, Ottava Rima, Gordon Ecker, Rhododendrites, Brews ohare, Cacadril, RexxS, JKeck, Against the current, SkyLined, Addbot, DOI bot, Lightbot, Skippy le Grand Gourou, Luckas-bot, Planlips, AnomieBOT, Jim1138, JackieBot, Citation bot, Bci2, ArthurBot, Xqbot, Neil95, Triclops200, Omnipaedista, TorKr, 乙乙, Paine Ellsworth, Ivoras, Citation bot 1, Pekayer11, Rameshngbot, PNG, RjwilmsiBot, TjBot, Lilcal89012, EmausBot, Socob, JSquish, StringTheory11, Quondum, TyA, Maschen, RolteVolte, ClueBot NG, Timothy jordan, Maplelanefarm, Bibcode Bot, BG19bot, Gravitoweak, Cadiomals, Tropcho, Fraulein451, DrHjmHam, Rhlozier, D.shinkaruk, Yaara dildaara, BronzeRatio, Monkbot, Yikkayaya, KasparBot and Anonymous: 142

- **Gluon field** *Source:* https://en.wikipedia.org/wiki/Gluon_field?oldid=669082789 *Contributors:* Rjwilmsi, Wavelength, Headbomb, R'n'B, Jonesey95, Arbnos, Maschen, Bibcode Bot, BG19bot, ChrisGualtieri and Anonymous: 2

- **Gluon field strength tensor** *Source:* https://en.wikipedia.org/wiki/Gluon_field_strength_tensor?oldid=677214274 *Contributors:* Rjwilmsi, Wavelength, Incnis Mrsi, Headbomb, Magioladitis, Connor Behan, R'n'B, Jacopo Werther, FrescoBot, Jonesey95, Ganondolf, Earthandmoon, Arbnos, Maschen, Bibcode Bot, BG19bot and Anonymous: 8

- **Color confinement** *Source:* https://en.wikipedia.org/wiki/Color_confinement?oldid=683221681 *Contributors:* Lorenzarius, Michael Hardy, Schneelocke, Timwi, Doradus, Phys, Secretlondon, Robbot, Ruakh, Isopropyl, Giftlite, Herbee, Xerxes314, Tagishsimon, Lumidek, MuDavid, Jag123, Cortonin, Jon.baldwin, Mpatel, Isnow, Ashmoo, Rjwilmsi, FlaBot, Pfctdayelise, Chobot, Mushin, Bambaiah, Hairy Dude, Ohwilleke, Salsb, E2mb0t~enwiki, Tetracube, WAS 4.250, Banus, SmackBot, Incnis Mrsi, Jjalexand, DHN-bot~enwiki, Sbharris, Colonies Chris, VMS Mosaic, Lambiam, Sinistrum, TaggedJC, Newone, Treue~enwiki, Thijs!bot, Headbomb, Cultural Freedom, Yill577, Natsirtguy, Melamed katz, A Nobody, Spoxjox, Anonymous Dissident, SieBot, FlowerFaerie087, Denisarona, Asher196, Manishearth, Dab240, Mustufailed, Texas Chainstore Manager, Addbot, Dudemanfellabra, Db1101, Eutactic, Luckas-bot, AnomieBOT, Materialscientist, Xqbot, D'ohBot, Jonesey95, RedBot, TobeBot, Trappist the monk, DixonDBot, EmausBot, Peaceray, Maschen, StanS, Helpful Pixie Bot, Bibcode Bot, Orentago, BG19bot, Slinkblot, Jdellamalva, Tikki and Anonymous: 48

- **Quark–gluon plasma** *Source:* https://en.wikipedia.org/wiki/Quark%E2%80%93gluon_plasma?oldid=681306263 *Contributors:* Taw, Michael Hardy, Cyde, Karada, SebastianHelm, Charles Matthews, David Newton, Grendelkhan, Phys, Dmytro, David Edgar, JerryFriedman, Art Carlson, Herbee, Rick Block, HorsePunchKid, Mako098765, Deglr6328, Squash, MuDavid, Ylai, Bender235, CheekyMonkey, Haxwell, Bradkittenbrink, Enric Naval, Cmdrjameson, Supercrisis, Jag123, Sam Korn, Fwb22, Anthony Appleyard, Axl, Hu, Knowledge Seeker, Cal 1234, Vuo, Joriki, Firsfron, Mpatel, GregorB, SDC, Palica, RichardWeiss, Ashmoo, Yuriybrisk, Maros, Ae77, Bubba73, Nihiltres, Goudzovski, Silversmith, YurikBot, Wavelength, Mushin, Bambaiah, Hairy Dude, Hellbus, Salsb, NawlinWiki, CecilWard, E2mb0t~enwiki, Curpsbot-unicodify, Ilmari Karonen, Ybbor, KasugaHuang, Neier, SmackBot, Stepa, PeterSymonds, Skizzik, Kmarinas86, Chris the speller, Silly rabbit, Jbergquist, Khazar, Dark Formal, Vampus, Fangfufu, JayHenry, Petr Matas, CmdrObot, Foice, Van helsing, Ruslik0, Michael C Price, Thijs!bot, Headbomb, Nick Number, Eb.eric, JAnDbot, Xeno, Yill577, Savant13, 28421u2232nfenfcenc, Ethron, MartinBot, Pagw, CommonsDelinker, J.delanoy, Maurice Carbonaro, Jeepday, DorganBot, 1812ahill, Momo Hemo, Fences and windows, BotKung, Pamputt, Ptrslv72, AlleborgoBot, Logan, SieBot, BotMultichill, Triwbe, Maelgwnbot, ClueBot, Flaming, Thunderhippo, Brews ohare, Mstrickl, Healyhatman, DumZiBoT, XLinkBot, Oldnoah, SkyLined, Truthnlove, Stormcloud51090, Addbot, Mjamja, Qmark42, Tide rolls, Lightbot, OlEnglish, מלמד בי, Luckas-bot, Yobot, Amirobot, 4th-otaku, AnomieBOT, Essin, Citation bot, ArthurBot, ProtectionTaggingBot, False vacuum, Ciceronibus, FrescoBot, Citation bot 1, Naxuesen, Tom.Reding, RedBot, Johann137, IVAN3MAN, Meier99, Puzl bustr, EmausBot, Mnkyman, Naznin farhah, ZéroBot, SalGiandinoto, Arbnos, Yiosie2356, SporkBot, Jesanj, Rangoon11, ClueBot NG, Jack Greenmaven, Raktimabir, Theopolisme, Helpful Pixie Bot, Bibcode Bot, 2001:db8, Shawn Worthington Laser Plasma, BattyBot, Kalmiopsiskid, Chemya, Saehry, Epicgenius, Prokaryotes, Polytope24, Pcharito, Vieque, Sofia Koutsouveli, KH-1, Crystallizedcarbon, Isambard Kingdom, Qulos and Anonymous: 115

- **Glueball** *Source:* https://en.wikipedia.org/wiki/Glueball?oldid=671829421 *Contributors:* Paul A, Loren Rosen, Phys, Sanders muc, Xerxes314, Mennonot, MuDavid, Jeodesic, Bambaiah, Hairy Dude, Ohwilleke, Xaxafrad, Smurrayinchester, Triple333, Saravask, Kmarinas86, V1adis1av, Sasata, Zaphody3k, Thijs!bot, Headbomb, Magioladitis, Nyq, Idioma-bot, Anonymous Dissident, Antixt, YonaBot, Avidallred, Boemmels, Alexbot, SchreiberBike, Addbot, Mpfiz, Lightbot, Luckas-bot, Dreamer08, AnomieBOT, Pra1998, Tom.Reding, Loqueelvientoajuarez, RjwilmsiBot, Carbosi, Drummermean, JSquish, ZéroBot, Suslindisambiguator, Whoop whoop pull up, Bibcode Bot, Vkpd11, Retnuh66, ChrisGualtieri, Richardbernstein and Anonymous: 10

- **List of mesons** *Source:* https://en.wikipedia.org/wiki/List_of_mesons?oldid=679943368 *Contributors:* Cherkash, Donarreiskoffer, Giftlite, Xerxes314, Michael Devore, Eequor, Rich Farmbrough, ZeroOne, Tompw, Physicistjedi, Pearle, Keenan Pepper, Zyqqh, TenOfAllTrades, Woohookitty, Ch'marr, Kbdank71, JVz, Strait, Nihiltres, Agerom, RussBot, David McCormick, SCZenz, Gadget850, Sbyrnes321, That Guy, From That Show!, SmackBot, JorisvS, Happy-melon, Charles Baynham, Chrumps, Usgnus, Cydebot, Christian75, Coccoinomane, Headbomb, Stannered, JAnDbot, Magioladitis, Mollwollfumble, Gwern, Leyo, Potatoswatter, VolkovBot, Antixt, Ocsenave, Muhends, Mikaey, SkyLined, Addbot, Yobot, Kan8eDie, 4th-otaku, Rubinbot, Citation bot, ArthurBot, Xqbot, Ulm, Carlog3, W-C, Yehoshua2, Citation bot 1, Thinking

8.3.2   Images

8.3.3   Content license